Television in Europe

Edited by James A. Coleman
and Brigitte Rollet

intellect™
Exeter, England

intellect

EUROPEAN STUDIES SERIES General Editor: *Keith Cameron*

Available now

Humour and History	Edited by Keith Cameron
The Nation: Myth or Reality?	Edited by Keith Cameron
Regionalism in Europe	Edited by Peter Wagstaff
Women in European Theatre	Edited by Elizabeth Woodrough
Children and Propaganda	Judith Proud
The New Russia	Edited by Michael Pursglove
English Language in Europe	Edited by Reinhard Hartmann
Food in European Literature	Edited by John Wilkins
Theatre and Europe	Christopher McCullough
European Identity in Cinema	Edited by Wendy Everett

Published in 1997/98:

Language, Community and the State	Dennis Ager
Gender in European Culture	Edited by Maya Slater
Policing in Europe	Bill Tupman

First Published in 1997 by
Intellect Books
EFAE, Earl Richards Road North, Exeter EX2 6AS, UK

Series editor: Keith Cameron
Production: Robin Beecroft
Copy editor: Kevin O'Brien
Cover illustration: Myles Hunt

British Library Cataloguing in Publication Data Available

ISBN 1-871516-92-7

Printed and bound in Great Britain by Cromwell Press, Wiltshire

Contents

Television in Europe: issues and developments

James A. Coleman and Brigitte Rollet

Television is inseparable from the development of Europe since the Second World War. The historical events unfolding in individual European countries have been witnessed by the television camera and made known, nationally and internationally, by television broadcasting. At the same time, the very existence of television has made a powerful contribution to the reshaping of society in individual countries as in Europe as a whole. Television can subvert a totalitarian regime by giving a vision of freedom in defiance of closed borders. Television can equally undermine indigenous regional and national cultures by proposing increasingly homogenised consumerist models for European viewers to aspire to. Television records and reflects society, but it also shapes society.

How does television work?

Television works by converting both sound waves and light waves into electronic signals. The picture, captured by a target plate within the camera, is scanned by a beam moving rapidly in successive parallel horizontal tracks from top to bottom, switching off between the end of one line and the start of the next. In colour television, such electron beams from three separate tubes register the intensity of red, green and blue light respectively. At the receiving end of the transmission, corresponding electron guns reconstitute the picture, which is recombined with the amplified audio signal. The number of lines scanned in for each image, and the number of scans per second (two interlaced scans make up a 'frame'), together determine the picture quality. Baird began with only 30 lines, but today there are three standards: PAL, originated in the UK and widely used in Europe, which has 625 lines and 25 frames per second; NTSC, the inferior American standard jocularly known as Never Twice the Same Colour; and SECAM, developed in France in line with General De Gaulle's policy of independence in all things, and adopted in friendly countries, as well as in the Soviet bloc as a result of a state visit by the General. In all three systems, the rapid succession of still images deceives the human eye and brain into perceiving actual movement. In some systems, a few of the lines may be used to transmit the equivalent of a primitive computer screen ('teletext').

The signal can be transmitted by wires (coaxial cable, fibre optic cable) or by electronic radiation (radio waves or microwaves). The video signal is composed of fluctuating voltages corresponding to fluctuating light intensities in each element of the picture, and made up of luminance (brightness) and chrominance (hue and colour saturation). It is accompanied by synchronising pulses guaranteeing correct reading by the receiver, and blanking pulses when the scanner is moving from one line to the next, together with an FM (frequency modulation) audio signal.

Because a terrestrial (or Hertzian) television broadcast requires a frequency band 400 times wider than that of a radio signal, the number of permitted stations in any location is severely limited. To squeeze in the largest number without interference means using very high and ultra high frequencies, despite the additional problems this incurs - shadows and echoes from buildings and other obstructions, and the shorter broadcasting distance, necessitating more frequent broadcasting antennae and directional, signal-boosting receiving aerials. Terrestrial television is therefore an essentially domestic medium, readily regulated by the political powers of the nation state within whose borders the transmitters stand. Satellites are another matter altogether, more efficient and far more difficult to regulate.

Satellites

Satellites are a means of receiving a signal, boosting it, and retransmitting it over a larger area ('footprint') than terrestrial means allow. The satellites are in geostationary orbit: in other words, they rotate with the earth, in the same position relative to the earth's surface, at a distance of 35,800 km above the equator, where the magnetic field (earth gravity) and centrifugal force are exactly matched. Broadcasters use satellites for the same reason as home-builders put lights on ceilings rather than on skirting boards – the signal is sent in all directions at once with minimal obstruction by other objects.

Whereas early satellites (Echo 1 was launched in 1960) merely received a signal from a fixed point (uplink) on the earth's surface and reflected it to another fixed point, today's multi-point to multi-point satellites can not only transmit to millions of domestic antennae (dishes) over a wide geographical area, but can also receive signals, direct or re-transmitted from any location on the earth. It was the development of small, portable satellite uplinks which allowed the world to watch the 1991 Gulf War live on television. Many satellites additionally have encryption facilities. Encryption is a means of scrambling a signal so that only certain receivers equipped with a decoder (normally, those who have paid a subscription) can receive it.

Analogue and digital

Until recently, sound and video recording and reproduction meant capturing the sound or image as it naturally occurs, i.e. as a continually varying signal. With this 'analogue' technology, quality can be downgraded by transmission, copying or in storage, a disadvantage obviated when the sound and picture information is held and manipulated in electronic form, in other words as digital data. This is why the world of television is moving towards digital technology, the technology already used for compact discs. To produce a music CD, mathematical values are ascribed to the frequency and intensity of sound waves by measuring ('sampling') them thousands of times every second. The values of each discrete snapshot in the series are expressed in binary form, and can thus be stored, transmitted or manipulated electronically, like any other computer data. In digital video, the technology is extended to recording the position, hue and intensity of each element of a picture. In each case the binary values ('bits') can be transmitted, by cable or radio waves, as electrical pulses, and the sound and picture reconstituted by the receiver.

The advent of digital video has been delayed because previous generations of computers lacked the speed to digitise ('compress') and reconvert to analogue ('decompress') twenty-five frames each second, even though the process is simplified by storing picture elements which remain the same and only recording afresh those which have changed from the previous frame. Using fewer frames per second and a very small (and therefore less detailed) picture made early digital video on CD-ROM feasible, but jerky and unsatisfactory. Advancing technology will shortly make full-screen, full-motion video widely available - and without the loss of picture quality experienced with analogue broadcasting. HDTV (high definition television), offering superb picture quality, will be widespread by the early years of the next century.

Domestic recording will shortly follow the same path from analogue to digital as has been taken by producers and broadcasters. Analogue videotape, originally reel-to-reel but domesticated in cassette form in the 1980s, will be replaced by digital video discs (DVD). To be launched in Europe in 1997, DVDs will be the same size as CDs, and the technology may take three years to pass through the read-only and WORM (write once, read many) stages before becoming infinitely re-recordable, superseding once and for all tape-based storage.

Financing television

There are three fundamental methods of financing a television channel: from public funds, from advertisers, and from subscribers. An annual television licence is one model widely adopted in Western Europe. Direct state funding might appear an attractive option to broadcasters, but it brings two major problems. The first is that state funding - effectively a television tax - is rarely unaccompanied by state interference: the state channel has very frequently been seen as an arm of government, and probably its most powerful propaganda weapon. The BBC in its early days was an honourable exception. The second drawback is that politicians hate to be seen to be increasing the level of taxes, and so tend to hold the level of television licences down below the level of inflation (although the introduction of new technologies in fact means that broadcasters' need increases above the level of inflation to remain competitive). The majority of public stations therefore need to augment state funding through commercial operations: sales of audio and video cassettes, international sales of programmes, merchandising or franchising of popular characters, programme-related periodicals. Because of viewer resistance to dubbing and sub-titling, especially in the USA and UK, the level of programme sales is highly dependent on the language in which they are made, i.e. the language of the domestic market. Export markets are hard to find unless productions, whether films or television programmes, are originated in English. National cultural references can also damage sales: hence the widespread adoption of anodyne or North American cultural norms even in European-made programmes.

In many cases, revenue from licence fee and supplementary sources remains insufficient, and some states, such as Spain, Portugal and Russia, have no licence fee at all. Public channels therefore have recourse to advertising, an option which brings with it formidable constraints on programming policy. Advertising had been for a century a staple support of the written media, and newspapers and magazines were among the many voices opposing and delaying the advent of television advertising. Although in the UK a commercial channel funded by advertising began as early as 1955 (cf. Italy 1957), other countries have resisted the introduction of advertising for far longer: France until 1968, Germany until 1984, others such as Sweden and Denmark into the late 1980s, and Austria's terrestrial broadcasters are still holding out. All countries impose restrictions on advertising, often concerning the timing and frequency of commercial breaks, and excluding certain products such as cigarettes, while often regulating political or religious advertising.

Advertising time on television is a negotiable commodity, whose price depends on the number and profile of viewers. Viewing figures therefore become crucial to financial viability, and channels vie with each other for top ratings. During *la guerre de l'audimat* (the ratings war) in France, the cost to channels of three commodities - popular films, big sporting occasions, and leading TV personalities - rocketed. Advertisers' concern with viewing figures also brings a more fundamental change: instead of making programmes whose aim is to get the potential viewer to switch on, television companies schedule programmes which the fewest number of viewers will actually switch off. Hence the accusation of lowest-common-denominator programming, which is so bland, unadventurous and undemanding that no-one can be put off.

Technology also influences advertising. Over the last decade, a remote control has often become standard television-viewing equipment. The word 'zapping' does not even figure in dictionaries of a decade ago, but today the practice is widespread and familiar to most viewers, so advertisers have had to make their productions even more eye-catching and attention-grabbing than before.

Sponsorship is a further method of attracting commercial funding for television, usually through the partial or total funding of a programme or series, in return for which the sponsors' logo or slogan is broadcast either side of commercial breaks. Product placement - in which a specific product is shown during a popular programme - can also bring in revenue.

For many years, there was scepticism concerning the viability of subscription channels. On the one hand, why should viewers pay additionally for programmes when they had free access already to several good channels broadcasting twenty-four hours a day? On the other hand, until recently, the technology allowing the supplier to encrypt (encode) broadcasts and to sell decoders in return for a subscription simply was not available. When the Sky satellite channel was launched in the UK, there was near unanimity among TV critics and financial experts that it could never be commercially viable. In fact, after several years of deficit trading and a merger with its only potential competitor BSB, and although the satellite diet of sport and films has not captured the whole market, BSkyB is now highly profitable.

The success of subscription channels must rest on distinctive programming, as Canal Plus, the first European subscription channel, demonstrated. Recent blockbuster movies and exclusive live coverage of major sports events are most often the key to attracting a paying audience, with occasional help from pornography which state channels are unable or unwilling to broadcast. Supplementary niche channels, devoted to news, sport, children's cartoons or foreign-language television, also help to make a subscription package attractive.

While the income of a television company has diverse sources, its outgoings, apart from administration and distribution, are concentrated on making or buying programmes. The choice between production and purchase is a crucial one for television station and viewer.

The increasing proportion of American-made programmes on European television screens can be ascribed to two causes - quality and cost. American series are professionally and expensively produced, with early piloting ensuring that the final programme will be popular. But with production costs met by the huge internal US market, the programmes can be offered to overseas buyers relatively cheaply. An episode of a top-flight bought-in American series can cost a tenth of a home-produced equivalent. Small wonder that schedules across Europe are dominated by US popular programmes, although national responses to the phenomenon range from untroubled acceptance to restrictive quotas designed to protect national languages and cultures.

With blockbuster drama recorded on film costing over £1m per hour, the expenditure can only be met if there is a huge internal market, as in the United States, or a ready international market, which is rarely available other than for English-language productions. Some countries seek to meet the shortfall by subsidies to producers.

History: common features, distinctive traditions

Although elements of the technology necessary to scan moving pictures had been developed by the German Paul Nipkow as early as 1884, the Scot John Logie Baird was the first, in London in 1926, to demonstrate what we would now recognise as television broadcasts. Experimental broadcasts in England began the same year, with a regular public service launched a decade later: in 1996, the UK celebrated television's sixtieth birthday. In France and Germany too, the mid-1930s saw the launch of regular television broadcasts, though reception was limited by restricted geographical coverage and by the cost of receivers.

Throughout the world, the Second World War interrupted the development of television services, but thereafter advances in electronic circuits and radio transmission and greater availability and affordability of components allowed television as a genuinely popular medium to develop. In the late 1940s in a few countries, in the 1950s in most, in the 1960s in Ireland and Greece, the state broadcasters set out to capture ever larger audiences. In the UK by the 1950s, and in Germany a decade later, a television could be found in the majority of households, and the threat to other media was clear. Families adopted the habit of watching the evening news: the information was fresher than in the press, and the use of moving pictures inevitably enhanced presentation, even if, then as now, television conveys emotions and impressions better than analysis. The expansion of television had an adverse impact as early as the 1950s on alternative entertainments, especially the cinema, with which the story of television has been tied ever since.

In most cases, television remained an arm of the state, administratively often a branch of the post office. When President Pompidou pronounced in 1969 that French television was '*la voix de la France*', he was asserting the right of the state government to exploit television broadcasts in its own political interest, both within and beyond national borders, as well as claiming for television a role in upholding national cultural and linguistic norms. At that date, and with the exception of the UK whose television was then freer of political influence, few in Europe, East or West, would have disagreed with his view. Indeed, the Irish Taoiseach's 1966 parliamentary statement on RTE (see Chapter 6) is in exactly the same vein.

In the last fifteen or twenty years, the majority of European governments, pressured both by public opinion and by the arrival of a multiplicity of internationally-sourced competitive channels, have sought explicitly to distance the public service broadcasters from direct political influence, to loosen the 'umbilical cord', but in many countries the regulatory body remains in an ambiguous position. And we must not forget that explicit political control is still possible in the 1990s, as evidenced by the banning of interviews with Sinn Fein on British and Irish television. International comparisons themselves have sometimes influenced national broadcasting policy, as in the case of the establishment of Italy's supposedly independent Guarantor for Broadcasting and Publishing. In all cases, the inevitable reliance on commercial funding has been accompanied by an equally inevitable loosening of the tight grip on broadcasting which virtually all governments enjoyed in the early days of radio and television, and which often persisted until very recently.

Because, on trips abroad, we see the odd evening's television and recognise the formats, the style of programmes, even a fair number of programmes themselves in dubbed or cloned form, we may tend to assume that television has followed the same developmental path in all European countries. This book shows how wrong such an impression would be. Nevertheless, there is today a degree of uniformity, and a number of common features do emerge from the individual chapters which follow. The history of television in Europe begins with a single national channel, administered, financed and heavily influenced by the state. From the 1960s on, there is typically a second public service channel, followed by one and then many commercial channels as the state monopoly is broken, although often, as for example in France and Russia, the habit of state control of a docile medium proves very hard to break, and everywhere there is tension between the role of public service broadcaster, with a mission to protect and promote the nation's cultural and linguistic heritage and audiovisual industry, and the role of entertainer which is necessary to boost viewing figures and thus attract advertising revenue as the cost of programming and technological progress outstrips the money governments are willing to raise from taxpayers. Local production is encouraged, either through local stations or through local slots on national stations. Colour broadcasts are progressively introduced from the early 1970s. Two newer technologies for distributing programmes become widespread - cable and satellite, either separately or in combination, with satellite, e.g. in Ireland, feeding local cable networks rather than individual dishes. At the same time, encryption makes financing feasible on a subscription or pay-per-view basis. In the 1990s, the number of available channels proliferates.

Programme concepts generally travel well. *Intervilles* grew into *Jeux sans frontières*; *Des chiffres et des lettres* was naturalised in the UK as *Countdown*. A British addict of *The Price is Right* can watch *Le juste prix* or *OK il prezzo è giusto* while abroad. In sport, in addition to world championships and European championships, national games are shown in other countries (e.g. the Italian football league *Serie A* on UK's Channel 4) and Europe-wide championships such as the soccer Champions League or more recently the Rugby Union European League are supplanting national leagues.

Although television schedules take into account the fact that the French tend to eat their evening meal later than the British and the Spanish later still, there are, then, recognisable similarities in programming across Europe, with a diet of sport, regular news broadcasts, occasional documentaries, films old and new, good and bad, a large quantity of variety shows, game shows, chat shows, audience participation shows, soaps and sitcoms, music videos, drama series, children's programmes, daytime tele-shopping, and cartoons. In an increasingly self-referential medium, there are chat shows and quizzes devoted to sport, and television game shows about television game shows. The use of less popular television slots for education, exemplified by Schools Broadcasting and the Open University in the UK, is parallelled in most other countries.

But similarities go beyond programming. With major sports events, recent blockbuster movies and top presenters in limited supply, prices rise, and new sources of income, including advertising and sponsorship, have to be found. The solutions adopted to meet financial problems tend to be shared, as are the difficulties of unravelling the complex international ownership of television channels and of transnational media empires, which may embrace newspaper, magazine and multimedia publishing and involve many levels of tie-ins and cross-medium promotion. The choice of digital decoder for BSkyB, to be made by Rupert Murdoch, will probably determine also the success of public terrestrial digital television, and illustrates the power now held by global media magnates.

Nevertheless, the case studies in this book also bring out distinctive features, such as the role of the Länder in Germany or of the army in Greece, or the way in which the geographical and political isolation of Ireland contrasted with the geographical centrality which gave Belgium early access to a range of channels. The cabling of Belgium began in 1960, and competition between the many channels it offered influenced the opening up of television a quarter of a century earlier than in, say, the UK or France, where cable is only now becoming available to most urban households.

No other country has experienced political change to match the upheaval which in the past decade has seen the USSR transformed into a number of independent republics. Like other industries, Russian television has shown the rapid adaptation needed to transform a monolithic state enterprise, raised on state control and meeting arbitrary targets, to the realities of a free market. In Russian TV, as throughout the Russian Federation, managers are having to refocus their attention from supply to demand. They must continue to please political masters, while developing the notion of a customer whose views matter. As in the rest of Europe, satisfying the advertisers by attracting the highest ratings has meant a diet of soaps, '*tok-shou*' and game shows. Russia has followed broadly the same path as western broadcasters, but has had to travel further and faster.

Television without frontiers

The thought of a multiplicity of television channels is still a dazzling prospect for the majority of UK households who do not yet have satellite or cable, but already a reality in many countries. Belgium and the Netherlands exemplify the permeability of internal and external boundaries, with a sophisticated and often polyglot audience enjoying the television output of several adjacent and more distant countries. At the same time as it embodies internationalism, the experience of Belgium epitomises the movement from the nation state towards *l'Europe des régions*, a tendency which technology facilitates by reducing the costs of local television distribution. The continuing success of the Flemish VTM also marks a further trend, the *fidélisation* of a local audience, while underlining the difficulty of predicting the impact of wider competition.

Terrestrial broadcasting has always reached across borders to threaten the monopoly of national channels. The United Kingdom is a set of islands, so that receiving broadcasts from adjacent countries, thanks to new technologies, is still a novelty for most. In virtually all other countries in Western Europe, however, at least in border areas, viewers have been able to eavesdrop on their neighbours for decades. West Germans had access to German-language broadcasts from the GDR, Switzerland and Austria twenty years ago. In the German Democratic Republic (GDR) in the 1980s, news and current affairs programmes from the Federal Republic (West Germany) offered an alternative view to state propaganda. Before the founding of Telefis Eireann, Irish households had no choice but to watch programmes from the country whose political domination they had escaped only forty years earlier, and whose economic domination was still a focus of resentment.

Technological developments have undermined still further the control of national governments over what their citizens watch. Satellites look down on a borderless world: from space, national frontiers are invisible, and national jurisdiction of television broadcasting is all but impossible. Satellite technology has been one major factor dictating Europe-wide measures (see below).

Meaningful access to other national and international broadcasters is one of many aspects of television in Europe where the language factor is significant. Language can be a double-edged sword. Because it naturally makes programmes in English, the BBC, unlike its French or Italian counterparts, is able to sell domestic products directly to a huge worldwide market. At the same time, the mass market in the UK, because of the vast, uninterrupted supply of programmes in a variety of Englishes from around the world, expects nothing less, and is highly resistant to dubbed or subtitled programmes other than feature films. The wave of imports does not threaten to overwhelm the BBC. But in Ireland, a far smaller state, possession of an English-language domestic market may be seen as an additional challenge which RTE, unlike RTBF or RAI, has to face. The predominance of English-language channels on the global airways is less of a threat to national broadcasters in countries whose mother tongue is not English.

A number of countries, including France, Greece and Ireland, have legislated to help protect and promote indigenous languages in the face of the onslaught from world English. Belgium, with its two national languages, provides a special case.

The European context

The broadcasting systems of most European countries have encountered massive changes during the 1980s. The introduction of privately funded commercial channels and the development of new transmission systems (cable and satellite) mean that European national television systems at the end of the century are far from what they were only a quarter of a century ago. The significant deregulation of European national mass media has made it more difficult for a European regulatory power to emerge. At the same time, it is towards the end of the 1980s that the idea of a 'European Audiovisual Policy' started to make sense and to give way to concrete common European projects. This does not mean, however, that the idea of a single European broadcasting system, or of transnational television was a new one.

European broadcasting: Eurikon (1982), Europa TV (1985), Eurosport (1989) and Euronews (1993)

The European Broadcasting Union (EBU) was created in 1950 as the pan-European organisation of public broadcasters, and a few years later the Eurovision system started. It was a non-political and non-commercial organisation which now has thirty-nine members in thirty-two countries, including some beyond

Europe. Eurovision programmes (usually international events such as the Olympic Games or world-wide sporting competitions) were initially exchanged free of charge. Since 1988, the EBU has accepted private channels, and this led to major controversy when Eurosport, a consortium of sixteen members of the EBU and of the private News International, was created in February 1989. As an EBU member, it benefited from EBU programmes, which some national TV stations considered to be unfair competition. The EBU has initiated several European audiovisual projects, some more successful than others. Eurikon was a 'limited experiment' (Humphreys, 1996: 258) launched in 1982. Of the fifteen official participating members, only five were actually involved in the project: West Germany (ARD), the Netherlands (NOS), Austria (ORF), Italy (RAI) and the UK (IBA). It was an opportunity offered to the EBU by the European Space Agency's OTS (Orbital Test Satellite). Each member was in charge of programming a mixture of news and entertainment for one week. After this initial experiment and the launch of the powerful satellite Olympus, the EBU was offered free use for three years. Europa TV, the name given to this project, started in October 1985. Five European countries, all members of the EBU, were involved in the consortium: West Germany (ARD), the Netherlands (NOS), Ireland (RTE), Italy (RAI) and Portugal (RTP). The channel was a failure as it stopped broadcasting after fourteen months (December 1986). The main obstacle was the difficulty of integrating the cultural factor in programmes aimed at extremely diverse audiences. The next EBU project was therefore of a more familiar nature for the organisation. Eurosport was devoted to sports only, a field no doubt less controversial and internationally more 'consensual' than other kinds of programmes.

This earlier lack of consensus explains the delay in launching the European channel Euronews. The pan-European newcomer began in 1993. 'Euronews is seen as a major contribution from the public service institutions to the European dimension in television, and its purpose is to demonstrate the indispensability of the public service concept' (Hjarvard, 1993: 81). Unlike Eurosports, there will therefore be no commercial parties involved in the project. Starting in 1989 with the idea of creating a European equivalent of CNN, the seven initial EBU members were planning to broadcast the new channel in five different languages. In 1992, there were thirteen broadcaster members. The budget of Euronews is financed by them (50%), the remaining costs being paid mainly by advertising and financial help from the countries involved.

European media policy: Television without Frontiers (1984–1989)

Initially a Green Paper entitled *Television without Frontiers*, the draft of the directive targeted 'the opening up of intra-Community frontiers for national television programmes'. Submitted to the Commission on 30 April 1986, the Council directive was meant, *inter alia*, to extend the Treaty of Rome regarding the free exchange and flow of European audiovisual products. The first article of the directive stated that: 'Without prejudice to Article 14 [related to advertising] and the provisions of Chapter V [related to copyright], Member States shall not restrict the reception and retransmissions to their territories of broadcasts from other Member States for reasons which fall within the fields coordinated by this Directive'. This article was reproduced in the Council Directive of 3 October 1989 (Article 2.2).

In addition to this major innovation in the history of European broadcasting, France initiated a quota for European productions. To protect the European cultural and linguistic heritage - and the European television industry - France initially wanted the member states to require their broadcasters to dedicate 60% of their programmes to European creations. The final Directive did not retain this requirement as such, although as shown in the chapter devoted to French television it was actually implemented in France, and similar concerns exist elsewhere in Europe, including in Russia. Article 4 stated that: 'Member States shall ensure, where practicable and by appropriate means, that broadcasters reserve for European work [...] a majority of their transmission time, excluding the time appointed to news, sports events, games, advertising and teletext services'. The controversy related to quotas went beyond the European Union's borders. The USA reacted in a very negative way to what was perceived as European protectionism. For the American government, the quotas 'would infringe the principles of the Uruguay Round of the General Agreements on Tariffs and Trade (GATT)' (Humphreys, 1996: 273). The issue became a very serious matter and a key topic during the GATT meeting which took place in Geneva in August 1990, and later during the GATT negotiations in 1994. More recently, the idea of quotas has been rejected by the members of the European Union themselves, together with the French-initiated project to expand support for the European film industry (see below).

European audiovisual programmes and the production industry: MEDIA (1986,1990), Eureka (1988), Eurimages (1989).

Since the mid-1980s, the European Union has been developing several projects aimed at helping the creation and diffusion of audiovisual works. In the light of previous paragraphs, it is worth mentioning that France, and more especially its Socialist President and Minister for Culture, was behind most of the ventures realised so far.

The MEDIA programme (*Mesures pour l'Encouragement et le Développement de l'Industrie Audiovisuelle*: Measures to Encourage the Development of the Audiovisual Industry) was initially launched by the European Commission in 1986, and agreed by its ministers in 1990. According to the proposal submitted to the Commission on 4 May 1990, it 'sets out to improve the environment of audiovisual businesses without directly intervening in production'. It is meant to 'strengthen [...] the national industries through the distribution of their products on a community scale to a potential audience of 320 million people [and to] establish [...] arrangements for cooperation between these industries based on increased production and distribution capacities on the international market'. In order to achieve such a goal, MEDIA guaranteed financial help to produce and distribute European creations, and provided training in production through its two programmes: EAVE (European Audiovisual Entrepreneurs) and MBS-MEDIA Business School. It received an initial budget of ECU 200 million for five years. This amount was insufficient for such an ambitious project (it was less than the ECU 268 million French budget for national cinema production), and the French attempt to have it doubled failed in 1995. The budget is currently ECU 310 million.

MEDIA was complemented by another audiovisual programme suggested by the French President François Mitterrand at the end of the European Council which took place in Rhodes in December 1988. The '*Eureka Audiovisuel*' as it was soon to be called, happened at the same time as the difficult negotiations related to the 1989 Directive on European Broadcasting (see above). In order to develop this project further, a meeting called '*Les assises de l'Audiovisuel*' was organised in Paris in September 1989. Twenty-six participating countries (including non-European Union ones) signed a joint declaration which stated that they agreed with the following objectives of the *Audiovisual Eureka* (2 October 1989): - the emergence of a more transparent and dynamic audiovisual market - the launching of actions and concrete cooperation projects of interest for the future of the European programme industry [...] - the widest possible distribution of European programmes [...] - the development and the widest possible diffusion of production from countries having a limited geographic or linguistic coverage in Europe - the promotion of European technologies [...] for the production and transmission of films and audiovisual programmes.

The fourth point related to non-European Union members is worth underlining as this aspect allowed the project to be generally better received than the previous ones, which had concerned European Union members only.

Eurimages is a European support fund established at the end of 1988. Its full title, 'The European Support Fund for the Co-production and Distribution of Creative Cinematographic and Audiovisual work', says more about its role within the Council of Europe. It helps to finance co-productions involving at least three European countries. Despite a smaller budget than MEDIA, Eurimages have been involved in almost fifty co-productions and have allocated a total of 122 million francs after two years of existence (Michel and Angoulvent, 1992: 14).

Conclusion on the European context

In the light of what precedes, it is obvious that although the projects have not all been successful, there have been repeated attempts by the European Union to try to develop a more general and common audiovisual policy.

The present book

Each chapter has been commissioned from a specialist author exclusively for *Television and Europe*. In order to make the book as useful as possible, all authors have addressed a number of common topics, while at the same time highlighting differences and bringing out the specific features of television in each individual country.

Each chapter provides a history of broadcasting with details of key legislation and regulatory frameworks. Each also explores financing and ownership, advertising, programme production, satellite and cable channels, links between television and government, relationship with cinema, and looks at the profile of each major channel - its mission, programming and audience. Tables cover key dates in television history, and the principal television channels, with details of funding and market share, and a thumbnail sketch of the type of programme carried. A table of principal organisations, institutions, and abbreviations (where appropriate) is followed by a bibliography.

The scope of the present book makes it impossible to cover every European country in detail. Instead, we have opted to include a chapter on each of the major countries, and a number of case studies of smaller countries, each of which has both distinctive features and features shared with other European television systems. While the principal focus is on EU countries, the book would not be complete without an account of perhaps the largest and fastest evolving European television network, that of the Russian Federation. As might be expected, there are contrasts between, say, Ireland and Russia, but although the television audience in the latter is fifty times that of the former, they do face many of the same problems in the process of transformation in which long-established traditional values are replaced by global, consumerist ones. Principal EU countries not covered by an individual chapter are considered briefly below.

Principal Countries of the European Union

In the 1970s, both Spain and Portugal saw repressive political regimes replaced by democracies. Both the Spanish and Portuguese television networks have seen massive changes since the beginning of the 1980s, and their evolution shows similarities.

Channel name	**Date of creation**	**Ownership and funding**	**Viewers**	**Profile**
TVE1	1956	State, advertising	52.4	News, fiction
TVE2	1965	State, advertising	19.3	Sports,fiction, documentaries
Antenna3	1989	Press groups 25%; Banks 35%; Advertising 10%	NA (limited broadcasting)	NA
Gestavision Telecinco	1990		NA (limited broadcasting)	55% Spanish programmes, 40% European programmes
Canal +	1990	Canal + 25% El Pais 25%	NA (limited broadcasting)	55% Spanish programmes, 40% European programmes

Spain

In Spain, from only two public channels at the beginning of the 1980s, there are now thirteen channels available. *Radio Television Espanola*, the first Spanish channel, began broadcasting in 1956 to only a few thousand TV sets in the Madrid area. The second channel was created in 1965 and was on air for only a few hours each day. Both channels were heavily state controlled as 'television was constituted as a state monopoly dependent on the government' (de Mateo and Corbella, 1992: 192). By 1982, the seventeen autonomous regions were authorised to own their own channels. The Basques were the first ones to create their own channel (they now possess two which broadcast around 150 hours weekly). The

Portugal

Channel name	**Date of creation**	**Ownership and funding**	**Profile**
RTP1	1957	State, advertising (licence fees stopped 1990)	Documentaries, fiction, news, light entertainment
RTP2	1958	idem	Documentaries, sports, light entertainment
TVI	1992	Catholic church 20% CLT 10%	NA
SIC	1992	Publishers, Link with Brazilian group Globo	NA

Channel name	Date of creation	Ownership and funding	Viewers	Profile: Programmes shared by associations:
Nederland 1	1953	Licence fees (40%), max. advertising: 26 minutes/day	20%	KRO (Catholic) NCRV (Protestant), VARA (socialist), EO (evangelical)
Nederland 2	1953	Licence fees, max. advertising: 20 minutes/day	19%	AVRO (liberal), TROS (populist), VERONICA (youth)
Nederland 3	1981	Licence fees, max. advertising: 5 minutes/day	12% (broadcasts 0500 -2300)	Cultural programmes, sports + 11 small associations
YRTLA	1990	CLT 40%, VNU 38%	29%	NA

The Netherlands

autonomous channels, which use their regional language, have created a consortium in order to promote and exchange their programmes. Commercial television in Spain only started in 1989. So far three private channels have been licensed to broadcast (including Canal Plus which use a similar system to Canal Plus in France, and which started in 1990).

Television in Portugal also started in the mid-1950s (1955). Both public channels (created respectively in 1958 and 1959) were strictly controlled by the state until the 1974 revolution, and reigned supreme until the Parliament agreed to break up the public sector monopoly and to open the audiovisual market to private channels in early 1991. Since 1990, the system of financing through licence fees has disappeared. Advertising will therefore remain the main source of funding.

The situation has been quite different for Northern European countries, especially the smaller ones which have had to protect their culture against massive imports of foreign programmes. Although Luxembourg does not have a national television as such, the CLT (*Compagnie Luxembourgeoise de Télévision*) which broadcasts in the country, has played a major role in European television, with a stake in many European countries, notably France, Belgium, and the Netherlands.

The system existing in the Netherlands is probably unique in Europe. It is based on the idea of 'segmented pluralism' through which any organisations, whether social, political or religious, can broadcast programmes on one of the three existing channels (created in 1953, 1968 and 1988). The number of programmes devoted to any of the organisations is in proportion to its importance within Dutch society. Always reluctant to accept commercial television, the government finally agreed in the mid-1980s to consider the deregulation of the

Denmark

Channel name	Date of creation/ privatisation	Ownership and funding	Viewers	Profile
TV1	1951	State	48% (1990)	Music, fiction, light entertainment
TV2	1988	State + advertising (maximum 10 minutes of broadcasting time)	52% (1990)	Fiction, light entertainment
Kanal 2	1988	Esselte 40%; Nordisk 30%; Others 30%	NA	NA

broadcasting network. Two channels circumvented the ban on commercial television thanks to the European directive 'Television without frontiers', and started broadcasting in 1989.

A similar reluctance and delay with regard to deregulating its broadcasting system can be found in Denmark. For decades, commercial television was not allowed at all, and the single channel, created in 1951, was financed only by the licence fee. A second channel, neither public nor private but autonomous and independent, started in 1988. Some of its broadcasting time is devoted to regional programmes. Local television stations are widespread in Denmark after the government started local broadcasting as an experiment in 1983. It permitted advertising in 1989.

'Newcomers' in the EU: Austria, Finland and Sweden

Broadcasting in the most recent European countries to join the European Union started - for different reasons - in the mid-1950s. The three countries share relatively small populations compared to most of their other European partners and a similarly limited number of public channels.

Austria

Following World War II and the division of Austria by the Allies into four sectors, the four victorious countries (France, the U.K, the U.S.A and the then U.S.S.R) took charge of the broadcasting network. It was only in 1955 that a national broadcasting institution was created (the ORF: *Osterreichische Rundfunk*). It was later (1957) to become a public service company. 'The first broadcasting legislation after the war came into force in 1967 and restructured the sector. Three radio channels and two television channels were defined as a minimum obligation for the ORF'. (Trappel, 1992: 3). No commercial television is allowed and even the satellite-channel 3SAT which Austria runs together with Germany and Switzerland is a non-commercial one. Most of the ORF income is generated by licence fees - the highest in Europe - and advertising. The Austrian government took part in some of the European projects listed above (MEDIA 92, Eureka and Eurimage).

Finland

The Finnish broadcasting company (YLE) was given a broadcasting monopoly as early as the Radio Act of 1927 (Tapper, 1992: 48). However, Finland started to develop its national broadcasting network at the same time as Austria. The first public channels (TV1 and TV2) started to broadcast together with the country's first commercial company (MTV Finland, 1957) which was allowed to transmit its programmes using TV1 and TV2. There are currently three Finnish public channels since the commercial channel TV3 was created in the mid-1980s.

Sweden

Although the Swedish Broadcasting Corporation was created in 1925, it was only in 1956 that television broadcasting started in Sweden. The second public channel was introduced in 1964. Advertising and sponsored programmes are not permitted on radio or television (Gustavsson, 1992: 215). It was only in 1987 that commercial stations started to transmit via satellite and cable (Filmnet, a pay-TV station, and TV3). Since then, another four commercial companies have begun broadcasting and one independent terrestrial channel has been allowed to introduce advertising. The development of the private sector has not affected the audience ratings for the public channels which are still far ahead of their private competitors.

Conclusion

Over the past seventy years, television has developed from a parochial, experimental technology to a highly significant economic activity and the most pervasive, universal component of contemporary culture. A typical European child spends more time watching television than in school. An adult spends a larger proportion of his or her life watching television than working. As in most other domains of contemporary life, the European television landscape is changing -and changing at an ever faster rate. From different starting points, and at different speeds, Europe's broadcasters have reached an increasingly homogeneous position, a position dictated by global consumerism. The common experience of broadcasters in recent decades might be characterised by the disappearance of state monopolies, deregulation, the growth of commercial stations and of local stations, increased television output, a financial crisis and competing scheduling pressures for public service broadcasters caught between the commercial realities and the desire to fulfil a national cultural and educational role, the growing power of trans-national players both in advertising and cross-media ownership. The globalisation of television parallels, documents and accentuates the globalisation of most aspects of contemporary life, and the correspondingly shrinking role of the nation state.

It is perhaps ironic that this book is structured according to nation states at a time when the nation state is being undermined, partly by the same forces which are moulding European television, and partly by television itself. The development of television in Europe may be seen as an inexorable move away from diversity and towards homogenisation. Standardisation has replaced the individual, sometimes idiosyncratic, national patterns of broadcasting, patterns which had grown up in an almost organic way to match the political, geographical, linguistic and social specificities of each individual country. As the political importance of the nation state decreases in Europe, many question whether the uniformity of European television will erode national identities. We have seen in the past international influences, including religious and political creeds, which have had to adapt to local circumstances; but now we face the opposite effect: television is unique in both its importance in individual lives, its centrality in social discourse, the speed of its diffusion, and its universality, now that the majority of the continent of Europe is saturated with television receivers. Television may bring unity at the price of uniformity. From now on, across Europe, the political, intellectual and cultural life of each one of us is inseparable from television.

Bibliography

Collins, R. (1992), *Satellite television in Western Europe*, John Libbey.

De Mateo, R., and Corsella, J.M. (1992), 'Spain' in *The Media in Western Europe*, Euromedia Research Group, pp.192-206.

European Community Policy in the Audiovisual field. Legal and political texts, Commission of the European Communities, 1990.

Gustavsson, K.E. (1992), 'Sweden' in *The Media in Western Europe*, Euromedia Research Group, pp.207-221.

Hjarvard, S. (1993), 'Pan-European Television News: Towards a European Political Public Sphere', in P. Drummond, R. Paterson and J. Willis (eds.), *National identity and Europe. The television revolution*, BFI, pp. 71-94.

Humphreys, P. (1996), *Mass media and media policy in Western Europe*, MUP.

Michel, H., and Angoulvent, A.-L. (1992), *Les télévisions en Europe*, PUF.

Proceedings of the European Audiovisual Conference, the European Commission, 1996.

Tapper, H. (1992), Finland' in *The Media in Western Europe*, Euromedia Research Group, pp.47-56.

'Télévisions en Europe' (1990), *Médiaspouvoirs*, no 20.

Trappel, J. (1992), Austria' in *The Media in Western Europe*, Euromedia Research Group, pp.1-15.

Television in the United Kingdom

Valerie Swales

According to paragraph 18 of the Home Office Committee's 1988 document, *The Future of Broadcasting*, 'Abroad British television is seen to be amongst the best, if not the best in the world. British broadcasters have set standards across the whole range of television programmes that we would wish to see maintained' (Stevenson and Smedley, 1989: 4). Yet currently it would seem to many that in the 1990s the deregulation of commercial television, the slimming down of the BBC and the expansion of cable and satellite transmission, which together have produced the most radical restructuring of British Broadcasting since the introduction of commercial television in 1954, seriously endanger this position. Taken together, the *Peacock Report* of 1986, the 1988 White Paper, *Broadcasting in the 1990s: Competition, Choice and Quality*, the 1990 Broadcasting Act, the 1992 White Paper, *The Future of the BBC: Serving the Nation, Competing World-Wide* and the 1995 Broadcasting Bill set out the agenda for television broadcasting in Britain in an increasingly competitive commercial environment where the provision of wider consumer choice is considered essential to success.

The Peacock Committee of 1986 set the agenda for much of what has come to pass in subsequent legislation. Its key recommendations were that there should be no advertising on BBC under the current structure of broadcasting, the licence fee should remain, 40% of programmes on BBC and ITV should be commissioned from independent producers within the next ten years, the selling of franchises should not be decided solely by the highest bids, Channel 4 should have the option to sell its own advertising, and that subscription should be used to secure consumer choice. The Home Office responded to these suggestions by arguing that the 'principles of public service broadcasting must be an integral part of the new broadcasting environment' (para 24), minority interests should continue to be served by established channels (para 34), there must be 'safeguards against concentration of ownership' (paras 42–3), market-led programming would not maintain diversity, information and education (para 91), the levy on ITV should be abolished (para 134), arrangements for the financing of Channel 4 should not change for the time being (para 139), the ethos of high quality television which has been the hallmark of BBC and Independent Television must be retained in some parts of the system (para 192) (Stevenson and Smedley, 1989: 18–21).

The government's 1988 White Paper, however, went against many of the key recommendations since its aims were to place the viewer at the centre of policy, with the expectation that more channels on air would mean greater choice, with viewers determining programming, either through subscription or through the ratings. The deregulation of ITV represented government intention to reduce the bargaining power of labour through both legislative acts and by forcing capital to reassess its (mainly labour) costs of production in response to Mrs Thatcher's view that commercial broadcasting was 'the last bastion of restrictive practices' (Dyson and Humphreys, 1990: 152). Regulation would work in the interests of decency rather than quality, and adequate standards would be achieved through competition. In addition, the independent production initiative, introduced in 1991 and made statutory in 1993, obliges both BBC and ITV to take at least 25% of their output from independent producers and encourages the utilisation of independent studios and production facilities. This last legislation, designed to create efficiency gains, has had a major impact on the structure of British television production.

History

To fully understand the nature of British Television it is necessary to consider the history of the BBC as a whole. Inaugurated in 1926, it became in 1927 the British Broadcasting Corporation. Having been granted a Royal Charter, it acquired a function of national significance beyond any mere commercial or entrepreneurial purpose. It was considered to have a unique organisation, unsuited to either private control or direct state management. In this respect it was a public corporation, conceived not as a profit-making institution but as a service available to the whole nation (Curran and Seaton, 1988: 119). Thus it became a new form of administrative entity, governed by social priorities. As a public broadcasting corporation it had a public monopoly status shared between politicians, civil servants and the people who ran it on a day to day basis. The first Board of Governors constituted a pattern of solid representation of the ruling class establishment, from high finance and business, the military, ex-public school, diplomatic service and foreign office. Equally, when commercial television was introduced in 1955, the members of the board of the Independent Television Authority came from similar backgrounds (Hood, 1983: 4). The first BBC Director General, Lord John Reith, who in the early days dominated the corporation, sought to impose through it a Christian morality on the British people. He immediately understood the power of the monopoly, believing it to be above politics. Indeed, the BBC was founded on the rejection of politics and evolved through the voice of the middle class liberal intelligentsia, providing guidance on matters of taste in music, correct pronunciation and cultural values. The role of television would be educative, would be to train character and, through a mixed programming policy, to provide entertainment for relaxation (Curran and Seaton, 1988: 161). Despite the non-political characterisation of the BBC, the procedure for the appointment of governors has been criticised for its secrecy. In particular, the appointment of Marmaduke Hussey in 1986 as Chair to the Board of Governors of the BBC was interpreted widely as a bid by the Conservative government to influence the management and running of the BBC, given his evidently close friendship with Mrs Thatcher and his previous position as Director of Times Newspapers Inc.

British television had commenced in 1936, and, after wartime closure, resumed broadcasting in 1946. Subsequently, the 1954 Television Act introduced a second independent network in direct competition to the BBC, thus creating the benign duopoly which was to characterise British television until the Broadcasting Act of 1990 signalled its end. The establishment of the Independent Television network in 1955 was to determine radically the shape of all television in so far as the BBC and ITV complemented each other as parts of a single public service as well as providing competition. However, the advent of commercial television was not won without considerable and passionate debate. Such factors as the dangers of a continuing monopoly, the unwieldy and increasing size of the BBC's bureaucracy, the threat from advertising to the public service ethos and the high standards and impartiality which had evolved over 25 years of broadcasting were augmented by press fears of threats to its own advertising revenue, anxieties about communist infiltration via purchased air time and distrust of American-style programming. However, the economic arguments in favour of commercial television were powerful. The first years of the 1950s saw a massive rise in consumption. The increase in sales of motor cars, television sets and telephones and the launch of branded detergents through extensive advertising campaigns left no one in doubt as to the profits to be derived from television advertising. The entertainments industries lobbied with Conservative backbench support for an independent commercial television network in the name of free enterprise and post-war prosperity. Ultimately the vast financial benefits to be gained from a new independent television sector had an economic logic which could not be ignored (Hood, 1983: 63).

ITV brought a new atmosphere to television since, run by entrepreneurs and entertainers, its motivations were to capture audiences, lower costs and make a profit through a formula closer to that of the USA: American soap operas, variety shows, quizzes, film series, chat shows and gameshows. By 1967 it was highly profitable with the airwaves, previously perceived as a publicly owned good, now a significant commodity for investment. By 1993 it had an aggregate revenue from advertising standing at £17,960 million, representing 28.5% of total advertising in Britain, and it currently enjoys 74% of the total television share of advertising, despite cable and satellite television now taking, at 6%, an increased share of the television advertising market.

Commercial television had a liberating effect as the BBC lost its establishment monopoly and popular taste finally found genuine representation (Curran and Seaton, 1988: 172). In the event, the editorial policy of Independent Television was outlined in accordance with that of the BBC, which meant that in Britain commercial television adopted the traditions of a public broadcast service. It was modelled to redress weaknesses in the BBC, to make broadcasting more sensitive to popular taste, to promote regional culture and oppose Londonism. Independent Television (ITV) was organised through franchised syndicates regulated by the Independent Television Authority (ITA) and was designed to promote competition not only between itself and the BBC but also between the ITV companies. In 1955 four major companies were set up, Associated-Rediffusion, Associated Television (ATV), ABC Television and Granada. In addition the ITA set up the Independent Television News (ITN) to provide national and worldwide news service under the companies' joint ownership. Competition was introduced by guaranteeing national access for agreed amounts of programming and by giving weekday contracts to one company and weekend contracts to another. Quality of content and quotas of programme type were controlled by the ITA/IBA (Independent Broadcasting Authority).

One of the main differences between BBC and ITV was intended to be the regional cover provided by ITV, although this was almost instantly negated by the institution of the network system and by 1968 the five large companies were obliged to produce network programmes (Seymour-Ure, 1992: 88). Further franchises followed, their economic base determined by the number of regional viewers they could serve, and by 1990 prior to deregulation fifteen companies plus ITN comprised commercial television. The anticipated outcome of distributing more widely the control of broadcasting did not ensue, but rather its opposite, with the amalgamation of the commercial franchises into fewer media empires.

The 1964 Television Act introduced a new model of direct taxation of ITV programme contractors, which came to be known as 'The Levy'. This was a payment beyond the rentals which the companies already made to the IBA and which the IBA was required to pay to the Treasury. The Act stipulated that the first £1.5 million of advertising receipts would not fall within this new taxation which was over and above normal taxation. Thereafter 25% of the next £6 million and 45% of anything over that would be paid. This proved too demanding for the television companies and subsequently a profit-based levy was introduced.

In 1964 a third channel, BBC 2, was launched, committed to providing quality broadcasting, especially drama, current affairs and documentaries. This was followed in 1982 by the introduction of Channel 4 to redress the balance of the duopoly. Born out of public service principles and ITV's expectation of the fourth channel, its function, complementing that of BBC 2, was to provide a national service for the social and cultural needs of so-called minority and specialist groups.

The 1990 Broadcasting Act heralded the end of the duopolistic arrangement which had characterised British television since 1954 by announcing the deregulation of the ITV network, the uncoupling of Channel 4 from its dependence on ITV and the launch of a new Channel 5. The auctioning of the new Channel 3 licences on December 31st 1992 via competitive tender was accompanied by a set levy and a projected end to protection from takeovers. Currently, Channel 3 comprises fifteen regional licences and one national breakfast service (GMTV). Other licensees awarded by ITC are Channel Four Television Corporation, Teletext Limited and two Commercial Additional Services. The funding formula for licence taxation, partly a fixed percentage and partly based on the original bids for tender, has caused widespread concern among the ITV companies who are currently lobbying to reduce their annual £373 million payment to the Treasury. The network has argued that it will find it increasingly difficult to make such payments and compete with competition from Channel 5, cable and satellite through reinvestment of profits in programming.

The Act also required that Channel 4 continue to appeal to tastes not normally catered for by ITV, that it remain distinctive and experimental, that it include educational provision and high quality news and current affairs programmes. More important, however, was the decision that the channel should no longer rely on advertising revenue received from the joint ITV companies but that it would now sell its own airtime independently while also making payment of a levy to the ITV companies as a safety net. The current Funding Formula was established by the Broadcasting Act of 1990 specifically to provide income support for Channel 4. Where Channel 4's revenue exceeds 14% of total terrestrial revenue, mainly from advertising and sponsorship income, the surplus has to be divided, with 50% being paid via the ITC to the Channel 3 licensees, 25% being paid into a reserve fund of the Channel 4 Corporation and the remaining 25% being available to Channel 4 for current expenditure. If Channel 4's qualifying revenue falls below the 14%, the shortfall must be made up from reserves. The ITC provided indicative figures in 1991 which determined that total terrestrial qualifying revenue was £1,993.6 million. Since Channel 4 had earned qualifying revenue of £393.9 million, £57.4 million had to be paid by the channel for distribution to the Channel 3 licensees and £28.7 million had to be placed in its reserve fund under statutory arrangements.

By 1998, under present arrangements, Channel 4 will have transferred £300 million to ITV. In this way Channel 4 has become the only broadcaster in the world with a public service remit to rely solely on commercial revenue with no public subsidy or licence fee. Also like BBC 1 and 2 the channel does not operate to make gains for shareholders but turns revenue into programmes on screen. Channel 4 has achieved its goals with resounding success, largely because of the decision to sell airtime at its real value. Net profit after tax and levy to the ITV companies stood at £128 million by December 1995, compared to £83.6 million in 1994. Yet the vexed question of its funding formula is still unresolved, although the Act allows the Secretary of State to amend the 14% formula, but not before the end of 1997. A government statement of intent indicated a two staged reduction in the funding formula leading to zero per cent in 1999 and certainly the government's proposal that the reserve be capped could produce £17 million per annum for reinvestment in programming and British film.

The third significant term of the 1990 Act was the announcement of a licence for a fifth and last free, conventional terrestrial service supported by advertising, with transmission projected for 1997. The tender was awarded to Channel 5 Broadcasting, a consortium led by the media companies Pearson and MAI. The service will operate on UHF channels 35 and 37, which means that it will only be semi-national, covering 70% of the country until cable and satellite is added. Furthermore channel interference has posed problems for aerial transmission and for video recorder output into the television set. Vertical polarisation of transmitter aerials is essential and will involve installation of aerials in millions of homes, whilst VCRs which have traditionally been tuned to Channel 36 are currently being retuned by an army of trained engineers. The new channel's programming strategy, broadcasting twenty-four hours a day, is to provide a simple schedule, assimilated at a glance. With a restricted budget of £110 million, the thrust will be very commercial. There is no intention to nurture rising talent, as does Channel 4, but to commission programmes from a small group of suppliers. It will be aimed at 35 year olds and is described broadly as a mix of Channel 4 elements with very popular sub-ITV material and strong on tabloid news.

Broadcasting Bill 1996

The Broadcasting Bill, published on 15 December 1995 as a substantial piece of legislation, laid down a framework for the expansion of both media ownership and viewer choice. The main points of the Bill were the following: the two-licence ownership limit on companies will be removed to be replaced by a limit of 15% share of total television audience; newspapers with more than 20% of national circulation will be barred from owning ITV licences; there will be a prospect of up to 36 digital channels with BBC given its own multiplex; the Channel 4 funding formula will be revised from 1998; there will be a merger of the Broadcasting Standards Council (BSC) and the Broadcasting Complaints Commission (BCC).

The decision to replace the limit on ownership has been welcomed by Channel 3 companies and newspaper groups who can control Channel 3 companies for the first time. Press holdings in television companies have long been an area of contention, although there is an apparently natural logic in one medium wishing to protect itself by investing in a competitor (Seymour-Ure, 1992: 92). The concern with regard to concentration of ownership and political pressure, however, has attracted scrutiny since legislation passed in 1963 and the Broadcasting Act of

1990 placed a limit of 20% on stakes that any newspaper might hold in television companies. The bar on Rupert Murdoch's empire which already controls more than 20% of total national newspaper circulation is certainly welcomed. However, Murdoch will be allowed to continue investment in cable and satellite and digital services.

The details of the framework mean that the BBC, with its own multiplex, will have the widest geographical cover, Channel 3, Channel 4 and S4C (Channel 4's Welsh language service) will share the second and Channel 5 will be offered half of the third multiplex. The government anticipates that the multiplexes will not be controlled by the broadcasters but rather by six large franchise providers with a twelve year licence, and there will be no payment to the Treasury for the first twelve years. From the consumer's perspective there have been criticisms of the government's paper which have concentrated largely on doubts about the consumer's willingness to bear the cost of new set-top decoders estimated at between £300 and £500.

The merger of the BSC and BCC, to be implemented on 1 April 1997, has been introduced as a solution to observations that confusion exists in the public's mind as to the role of the two regulatory bodies. Since 1990 (see below) BSC has been responsible for monitoring issues of sex, violence, bad language and decency, while the remit for BCC has been unfair treatment and infringements of privacy. However, there has been concern by both bodies at the apparent inconsistency in the upholding of complaints and it is hoped that a single committee will be able to offer rationalisation of outcomes.

Penetration and ownership

Screen-based entertainment in the home has increased sharply since the mid 1980s as the television screen has ceased to be merely a means of viewing terrestrial broadcasting output. Television ownership reached saturation point many years ago with fewer than 5% of homes being without one set since the mid 1970s. 80% of all households had a video recorder by 1993 whilst the use of teletext increased to 49%. Yet live terrestrial television viewing has remained remarkably steady over the last decade in Britain with weekly figures of virtually 100% for BBC 1 and ITV, 87% for BBC 2 and Channel 4 and 14% for other channels. The technological change needed to transform broadcasting by creating an 'open digital environment' has been facilitated by the laying of a national fibre optic grid and so far 6.5 million homes have been 'passed by'. The first satellite broadcast in Britain occurred in 1981 transmitted by Satellite Television, bought out by Rupert Murdoch two years later and renamed Sky Television. Cable soon followed, although take-up of satellite and cable have been slow in Britain, with subscribers to Sky currently standing at 4.2 million and reaching 13.2 million people in 20% of homes whilst cable has been taken up by 21% of homes. The majority of customers sign up to cable for the cheaper phone services rather than to buy into entertainment packages, preferring access to Murdoch's satellite channels. Major drawbacks for cable are low audience reach, indifferent standards, and magazine-style niche marketing, all of which fail to attract advertisers. Currently BSkyB is seen to have an effective market monopoly on pay television but, despite channel innovation, the main goal is still dish penetration which is driven by sport and movies, the premium services. Entertainment is expensive and with a budget of £35 million, Sky One is unable to compete with either ITV, with its budget of £850 million, or BBC with £1 billion.

Cable and satellite broadcasting

The government's digital broadcasting policy indicates increasing relaxation of cross channel ownership and states that the introduction of digital terrestrial television to Britain offers the only guarantee of universally available free-to-air national radio and television services. At September 1996 the cable and satellite programme channels listed by the ITC number 82 licensable programme services with one third non-operational and 130 non-domestic satellite services. Currently licensed satellite television services can be divided into those channels which cater to wide interests and those which provide specialist and minority programming. The former include general entertainment (Sky One, UK Gold, Adult Channel), feature films (Sky Movies, Movie Channel, Sky Movies Gold), sport (Sky Sports), music (MTV Europe, Cable Jukebox). Specialist channels offer programmes for children (Disney, Nickleodeon), and news (Sky News), for a wide range of ethnic communities, and increasingly viewers have access to programmes from overseas (Eurosport, CNN News).

The government recognises that there can be no accurate prediction of the take-up of cable, satellite telecommunications and terrestrial delivery systems and that there can be no guarantee of commercial success. It believes incentives must be created for households to invest in the new technological systems, although there are doubts about the consumer's willingness to bear the cost of new set-top decoders. Analogue broadcasting will continue alongside simulcasting with the possible introduction of between eighteen and thirty-six digital terrestrial channels, although it has been suggested that an end date for analogue television should be fixed to ensure the successful take-up of digital terrestrial television.

Regulatory framework

Both the BBC and commercial television are supervised by their own regulatory mechanisms, although there is much overlap between the two. In particular there is mutual agreement on the scheduling of programmes for family viewing from 6 o'clock in the evening and a nine o'clock watershed for more adult programming. The Governors of the BBC, its General Advisory Council, regional councils, and specialist advisory bodies for specific subjects are appointed to act as trustees of the public interest according to approved codes of practice. In fact the Governors perform two functions, facing, as they do, to corporation as public service and to government (Seymour-Ure, 1992: 63). In its turn the Independent Television Commission (ITC), with a similar structure, regulates commercial television, although with a much lighter touch. The ITC does not review programmes but rather acts as a monitoring body within the framework of its own Code acting on the assumption that licencees themselves will decide on the suitability of programmes. It has however stipulated that the regional Channel 3 companies are required to operate a system of networking throughout the UK with subtitling for the deaf of at least 50% of all programmes by 1998.

Although cable and satellite programmes are regulated by the ITC, no positive requirements were put in place to guarantee particular qualities or programme range. Moreover, rationing of licences was not introduced, implying that assumptions concerning the constraints imposed by terrestrial frequencies would not apply to multichannel cable or satellite. Consequently regulation is confined to the prevention of abuse in the form of unacceptable programming and advertising. The ITC's regulatory codes apply to services uplinked from the UK while it maintains a monitoring role for programmes transmitted from elsewhere.

Dual watersheds operate for cable and satellite with 8pm set as the limit for some programmes which otherwise might be scheduled after 9pm on terrestrial television, whilst 10pm has been set for the transmission of adult material. This departure reflects both the classification of video and the recognition of a different relationship between broadcaster and viewer where the latter is believed to make an independent and conscious choice about viewing. Equally, no clear policy as part of the Broadcasting Act exists for fixing the proportions of European-made programmes to be included in cable or satellite services, although the ITC worked with the Department of the National Heritage to produce the EU Directive on *Television Without Frontiers* which deals with transfrontier satellite broadcasting.

The development of new media has called for radical rethinking of communications in Britain and yet it is apparent that issues related to Europe and the wider internationalisation of television trade have not been addressed. Policies in Britain are still dominated by domestic interests and its cultural and linguistic isolation in Europe has allowed it to develop its broadcasting structures to suit its own needs. Rarely has it had to consider infringements of its 'cultural barriers and its consequent strength as a programming centre has enabled it to move outwards so exploiting the demand for high quality English-language television content' (Dyson and Humphreys, 1990: 150). There is little concern for the presence of America which has been long accepted, or for foreign programme quotas, and moves towards policies of protectionism are considered outdated.

In the Broadcasting Act of 1990, specific roles were outlined for the Broadcasting Standards Council (BSC) and the Broadcasting Complaints Commission (BCC) which have a remit to monitor incidents of 'unjust or unfair treatment' and invasions of privacy and issues of sex, violence, bad language, impartiality and inconvenient scheduling. The BCC was set up in 1981 with an independent chairman as a result of severe criticism by the Annan Committee of the lack of true accountability in both the BBC and ITV (Seymour-Ure, 1992: 238). The regulatory mechanisms for accountability which vouchsafe standards in British television, although seemingly ad hoc, offer in some respects an ideal for broadcasters, although fears have been voiced of a tendency towards excessive self-censorship in the name of decency. Regulation is currently based on arms length regulation by government, with the proviso that in a national emergency the government might intervene directly, the most notable example of this being the ban, now lifted, on the broadcasting of direct statements by members of Sinn Fein, a rule circumvented by broadcasters, who instead had the words delivered by actors' voices.

The ITC is also responsible for publishing and applying codes and rules on the content, amount and distribution of advertising and programme sponsorship. It reviews the rules, advising licencees about interpretation and monitoring and investigates complaints from viewers. It also controls advertising sales to ensure fair competition. The Advertising Advisory Committee assists the ITC in setting standards for television advertising and constantly reviews issues concerning such products as alcohol, sanitary protection and pharmaceuticals. In 1990 the Broadcasting Act lifted the statutory ban on sponsorship on terrestrial channels and by 1995 36 network programmes received sponsorship with a total revenue of £21 million. Sponsorship of cable and satellite continues to increase and is an established feature of dedicated satellite children's channels. Research showed that in 1995 there were no significant breaches of the Sponsorship Code.

Channel name	BBC1	BBC 2	Channel 3	Channel 4	Sky TV
Date of creation	1936	1963	1954, deregulated 1993	1982, deregulated 1991	1983
Ownership and funding	Public service: British Broadcasting Corporation; funded by licence fee	Public service: British broadcasting Corporation; funded by licence fee	Private, regional licencees, owned by conglomerates, eg MAI, Carlton, Pearson; funded mainly by advertising	Private, Channel Four Television Corporation; funded mainly by advertising	Private, owned by Rupert Murdoch; funded by subscription and advertising
Viewers	34%	10%	42%	11%	4%
Profile	Popular: news, features, documentaries, current affairs, sport, light entertainment, children's, acquired feature films, drama	High quality, features, documentaries, current affairs, news, sport, schools and continuing education, Open University	Popular: drama series, news, current affairs, sport, documentaries, serials, feature films, television films	Specialist interest: drama, sport, features, arts and music, film, foreign film, acquired series, light entertainment	Subscription; movies, sport, repeat television programmes

he Channels

Systems for regulation and accountability have been supplemented by extensive audience research since 1980 when a single system, the Broadcasters' Association Research Board (BARB), owned jointly by the BBC and ITV, formed to provide data information on viewing patterns. This data, the ratings, is compiled from 4,500 households by comparing viewing audience to the population as a whole and is demanded by shareholders and subscribers in the advertising and broadcasting industries and by the broadcasters themselves. Television Audience Measurement is a qualitative methodology which uses a 'peoplemeter', installed between 1984 and 1986 to deliver accurate, computer linked information on viewing of terrestrial, satellite and cable television and video playback. The Audience Research Service supplies data on audience reaction from a diary panel of households to the broadcasters only. In addition the Establishment Survey of Television Hours will conduct 44,000 interviews on a rolling basis throughout the year according to an agreed geographical spread. Audience response to programming is made public through broadcast programmes ranging from those offering genuine debate to those providing entertainment. *Right to Reply* (Channel 4) and *Bite Back* (BBC1) submit producers to public interrogation, whilst *Points of View* (BBC1) presents viewers comments in a light-hearted Sunday evening slot.

BBC 1 and BBC 2

The BBC has been obliged to seek a new identity as a public service broadcaster in anticipation of the renewal of its Charter in 1996 but in the knowledge that, according to the White Paper, *The Future of the BBC: Serving the Nation, Competing World-Wide*, it would continue to receive its chief source of revenue from the licence fee until 2001, would not be required to take advertising or sponsorship and would be encouraged to develop commercial broadcasting subsidies (Eckstein, 1994: 2). In 1992 the paper Extending Choice announced the BBC's 'Producer Choice' policy, whereby 25% of programmes must be purchased from independent companies. Also there would be an increase from one fifth to one third of network production transferred to the regions, with headquarters in Cardiff, Belfast,

Glasgow, Manchester, Birmingham and Bristol. The policy is designed to promote greater efficiency and increase competition as part of a greater plan to modernise the BBC's lumbering bureaucratic organisation. It is clear that production costs within the BBC have been lowered through the slimming down operation but the reduction has involved the shedding of jobs and, it is claimed, a decline in quality, a sacrificing of the creative freedom of programme makers to increasing bureaucracy and accountability, and uncertainty of employment for the smaller independent producer.

The principle of a flat rate licence fee, set and periodically reviewed by the government, and currently £89.50 for colour, has supported the notion of national service interest whilst providing the government with political leverage. At the same time, however, the unlikelihood that public television could always receive its entire funding from the licence fee was recognised as early as the 1930s, given the high expense incurred by television production. As market fragmentation and niche transmission increasingly shape the landscape of television, the principle of a public broadcasting service which provides for the needs of the majority of the population across the spectrum of television viewing comes into doubt. Consequently there has been much conjecture about the future role of the BBC as a programme provider.

Currently the BBC continues to appeal to all tastes and interests, embracing information, education and entertainment. A total output of BBC network broadcast hours of 12,806 in 1992/1993 represented weekly output hours of 124 hours for BBC1 and 122 hours for BBC 2. Overall news accounted for 10.4%, features, documentaries and current affairs represented a weekly output of 19.0%, drama output accounted for 4.2% and sport for 9.7% whilst there was a decrease in feature films and series from 22.0 to 19.6%. Taken separately, BBC 1 devotes more time to light entertainment and acquired series as well as to news and news related programmes. It also takes the major share of children's television. BBC 2 broadcasts more features, documentaries and current affairs, acquired feature films and nearly all education and Open University programmes (Eckstein, 1994: 20-21). Beyond its role as broadcaster, it takes seriously its responsibility as a major patron of the arts, supporting five orchestras and providing work for actors, musicians and composers.

The BBC is also an international broadcaster, setting up the World Service Television in 1991. It has also been encouraged to become an entrepreneur, and via BBC Enterprises Ltd sells merchandise such as programmes, video and audio cassettes, and publishes books and magazines. Trade in television programmes is vital to the British economy and allows Britain to maintain its reputation as the world's second most powerful television trading industry, with BBC Enterprises Ltd. the world's largest exporter in the field.

The cost to the BBC network of different programme categories is of necessity a significant factor in programme planning with outside live sports broadcasts costing upwards of £40,000 per hour, and drama well over £500,000 per hour. In other words, in 1992/1993 drama counted for 7% of total output but 29% of total costs whereas sport accounted for 31% of output at 8% of total costs. The response to demands to become increasingly cost effective has been to control the mix between first-time transmission and repeats. In 1992/1993 BBC 1 had fewer repeats than BBC 2 at 72.6% against 51.6%. First time transmission was largely accounted for by features, documentaries and current affairs and there is now a perceptible increase in award-winning programmes in these categories.

Light entertainment, drama and feature films accounted for only 16.4% of first-time transmission on BBC 1 and 13.0% on BBC 2. Acquired feature films and series made up 47.8% of repeats on BBC 1 and 32.8% on BBC 2 (Eckstein, 1994: 17).

The terms of renewal of the BBC's Charter in 1996 for a further ten year period signals a coup which underlines its current favourable position in relation to the government. Key changes between the old and the new charter lie in the statement of objectives which allows the BBC for the first time to be both a commercial and a public service broadcaster. The latter will remain the core activity and public funds will not be used for commercial activities. It is given the right to show BBC repeats on satellite services and to provide services either alone or with a partner which are funded by advertising, sponsorship, pay-per-view, subscription or any other means of finance. This development was already heralded by the launch of UK Gold. The holdings of UK Gold and subsequent ownership agreements clearly point the way to future collaboration between public service and commercial broadcasting in the bid for increasing market share. UK Gold, a highly profitable pay-television channel, transmitted by BSkyB, also holds the rights to the extensive library of the BBC. Pearson and Cox Communications, a US Cable Company, who both hold 15% stakes in UK Gold, are planning to sell their stakes to Flextech, the cable and satellite broadcaster controlled by the giant US cable firm ITC. This agreement paves the way for Flextech to complete its joint venture with the BBC to launch up to eight channels from Summer 1997. The aim is to make sure the BBC has maximum freedom to move into the digital age and to operate within the contexts of both licence fee and subscription, thereby avoiding long-term decline in both share of income and audience impact (Brown, 1995: 16).

Channel 3

Channel 3 currently operates in a climate of extreme uncertainty with incessant takeover speculation and an obligation to deliver to shareholders. In general terms, Channel 3 is committed to the development of sustainable revenue by building programme sales and by buying and exploiting programme rights, through national and international alliances. It favours the acquisition of media rights for products such as children's programmes, natural history and feature films, all of which are capable of wide distribution and have a long life with the potential for relicensing. Consequently, in recent years risk has been low and the programme mix has been relatively stable, while the awarding of franchises has had little impact on general programming. However, advertising is its principle source of revenue, and responding to falling figures in the audience ratings and criticism of programme quality from advertisers has been an impetus for the commissioning of a run of successful drama series such as *Band of Gold* and for the bringing forward of new programme schedules. In 1992/1993 news, magazines, documentaries and current affairs accounted for 24.9% of output with plays, drama, serials, feature films and television movies accounting for 34.0% (Eckstein, 1994: 18).

The ITC Programme Code has laid down guidelines for programming regarding taste and decency, impartiality in news programmes and controversial issues, prohibition of commercial promotion within programmes, and of brief images or other technical devices which might be broadcast without the viewer's being aware of them. However, the Code does not offer a comprehensive set of rules for all circumstances in all programmes. Channel 3 must meet significant strands of programme requirements with minimum time allocations for education, religion, art and children, must serve particular regional interests and broadcast a minimum quota of European programmes. Licensees must produce or commission a minimum of 65% of original programmes, including repeats, with diverse appeal. In 1995 independent production accounted for 26% of total ITV expenditure, representing 15% of all transmission time.

Channel 4

Upon deregulation in 1991 Channel 4 gained independence from Channel 3 and was required to operate its own advertising revenues. Against expectation, it has been most successful in attracting advertising which delivers over 95% of the channel's revenue. In part this is due to its commitment to building audiences biased towards upmarket, younger and lighter viewers. Also the channel is not required to deliver dividends to shareholders. This financially beneficial situation is currently under threat from suggestions of privatisation by the government. However, critics have observed that Channel 4 no longer satisfies its original mission to provide for specialist and minority interests and to encourage innovation and experiment, particularly if the large number of imports of American comedy series and foreign sport are considered. Indeed the policy to build audience share can now be seen to have had an impact on its programming. Between 1992 and 1993 entertainment rose from 25.5% to 31.0%, quizzes from 2.8% to 3.5%. Feature films decreased from 17.1% to 16.4%, news output dropped from 7.7% of output to 3.8% whilst documentaries and current affairs fell from 10.4% to 9.8%. Arts and music fell from 4.0% to 2.3% (Eckstein, 1994: 20-21).

Yet its programming has shown an increase in audience share of total live television from 10.0% in 1992 to 10.9% in 1995. Its ability to deliver audiences to the advertisers registers in the figure of £330 million from advertising revenues compared with £251 million subscription from ITV in 1992. Overall programme output from Channel 4 possibly does validate the claim that it no longer honours its remit given that a greater proportion of its output is devoted to entertainment than is the case with Channel 3. One conclusion to be drawn from this is of course that market forces have won the day. Nevertheless, the Channel still reserves a space for mordant, political satire with *Rory Bremner, Who Else?*, and still includes programmes of specialist interest such as *The Real McCoy*, which is widely popular and provides space especially for Asian and Afro-Caribbean comedy; *The Girlie Show* and *Jo Brand Through the Cakehole*, for and about women; *Out on Tuesday*, for gay viewers; foreign sport and film, and trenchant social commentary. Due to its financial success Channel 4 has partially won its case to keep more of its profits which it will use to broadcast fewer repeats, cut American imports, invest more in drama and extend its support of the British film industry with productions such as *Four Weddings and a Funeral*, *The Madness of King George*, and *Shallow Grave*.

Conclusion

As a major provider the British television industry has been forced to modernise in response to the vast expansion of the digital telecommunications system. Currently it is characterised by the rush for domination of commercial terrestrial television. Domestic and transnational media ownership as a fallout from deregulation, the global market place, the function and future of the BBC and public broadcasting, the emergence of a significant independent production sector, the sustainability of terrestrial television have all become key issues. What was understood as television in the 1960s and is still regarded as such by the majority of domestic viewers might well become only one aspect of the great wave of information which will enter the home with the vast expansion of the new media. Clearly, the new media landscape will drive a change which will have a major impact on the special status of public service broadcasting in Britain as it is placed back into the mainstream economy. Former public service ideals which by and large have driven issues of regulation and accountability will give way to competition, efficiency and consumer demand. Such developments will increasingly threaten high standards of programming, wide access, provision for minority interests and impartiality. Also they will produce inequalities in television viewing, not the least of which will be that those who are able to pay will receive the better service and greater choice. Britain may wish to be a major global player but it is firstly a domestic provider which trades in the European and international markets and must survive at home (Dyson and Humphreys, 1990: 152).

Key Dates

1936 British television begins.
1945 Broadcasting resumes after wartime closure.
1954 Television Act establishes ITV.
1955 Independent television commences.
1956 Eden makes first televised ministerial broadcast.
1957 Queen's Christmas message broadcast on TV.
1959 First general election coverage on television.
1964 BBC 2 commences.
1965 Cigarette advertising banned from television.
1967 BBC 2 starts regular colour.
ITV re-franchises.
1969 BBC 1 and ITV start regular colour.
1974 Annan Committee on broadcasting (reports 1977).
1981 First satellite transmission by Satellite Television.
1982 Channel 4 starts.
ITV re-franchises.
Hunt report signals cable expansion.
1983 Rupert Murdoch launches Sky Television.
Breakfast time television (BBC and ITV) begins.
1985 Peacock Committee on BBC finance (reports 1986).
1988 White Paper, *Broadcasting in the 90s: Competition, Choice and Quality*.
1990 Deregulation of ITV (Channel 3 commences 1991).
1992 White Paper, *The Future of the BBC: Serving the Nation, Competing World-wide*.
1996 Broadcasting Bill.
Renewal of BBC charter

Bibliography

BFI Film and Television Handbook 1995 (1994) London, BFI.
Brown, M. (1995) 'The BBC: back after the break?', *The Guardian*, 4 December 1995.
Corner, J. (1955) *Television Form and Public Address*, London, Edward Arnold.
Curran, J. and Gurevitch, M. (eds.) *Mass Media and Society*, London, Edward Arnold.
Curran, J. Gurevitch, M. and Woollacott (eds.) (1977) *Mass Communication and Society*, London, Edward Arnold.
Curran, J. and Seaton, J. (eds.) (1988) *Power Without Responsibility: The Press and Broadcasting in Britain*, London, Routledge.
Dahlgren, P. (1995) *Television and the Public Sphere: Citizenship, Democracy and the Media*, London, Sage Publications Ltd.
Dyson, K and Humphreys, P. (eds.) (1990), *The Political Economy of Communications*, London, Routledge.
Dyson, K. and Humphreys, P. with Negrine, R. and Simon, J.-P. (eds.) (1988) *Broadcasting and New Medias Policies in Western Europe: A Comparative Study of Technological Change and Public Policy*, London, Routledge.
Eckstein, J. (1994) *Cultural Trends*, Issue 21, London, Policy Studies Institute.
Ellis, J. (1982) *Visible Fictions*, London, Routledge.
Fiske, J. and Hartley J. (1978) *Reading Television*, London, Methuen.
Garnham, N. (1990) *Capitalism and Communication: Global Culture and the Economics of Information*. London: Sage.
Harvey, S. (1989) 'Deregulation, Innovation and Channel Four', *Screen*, 30, 60–78.
HMSO (1988) *Broadcasting in the 90s: Competition, Choice and Quality*.
HMSO (1992) *The Future of the BBC*.
Hood, S. (1983) *On Television*, London, Pluto Press.
Kuhn, R. (ed.) (1985) *The Politics of Broadcasting*, Beckenham, Croom Helm Ltd.
Macdonald, B. (1993) *Broadcasting the United Kingdom: A Guide to Information Sources*, London, Mansell Publishing Ltd.
Negrine, R. (1990) 'British Television in and Age of Change' in K. Dyson and P. Humphreys (eds.) *The Political Economy of Communications*, London, Routledge.
Paterson, R. (ed.) (1990) *Organising for Change*, London, British Film Institute.
Reith, J. (1924) *Broadcast over Britain*. London, Hodder and Stoughton,
Scannell, P. and Cardiff, D. (1991) *The Social History of British Broadcasting*. Volume One, Oxford, Blackwell.
Sendall, B. (1982) *Independent Television in Britain Vol. 1: origin and foundation 1946-62*, London, Macmillan.
Seymour-Ure, C. (1992) *The British Press since 1945*, Oxford, Blackwell,
Stevenson, W. and Smedley, N. (eds.) (1989) *Responses to the White Paper*, London, British Film Institute.

Television in France
Brigitte Rollet

History

French television started early: the first television programme was broadcast in 1935 from the Ministry of Telecommunications (or PTT). But compared with the situation in other industrialised countries, France was very slow in developing its terrestrial (Hertzian) network (see *Key Dates*). This could be partly explained by the importance of another medium during the inter-war period. Radio was one of the major broadcasting media and 'enjoyed a monopoly in the broadcasting field' (Kuhn, 1995: 77).

The first television experiment in France took place a few years earlier, in 1931, when a demonstration was held at the *Ecole Supérieure d'Electricité* at Malakoff near Paris. Despite a few regular programmes from 1937, the beginning of the war put a temporary end to French broadcasting. During the Occupation of France (1940-1944/5), the Germans launched Paris TV and broadcast recorded programmes. At the time, there were only 1000 television sets in Paris. After the war, programmes became more regular on the single black and white channel: they included the *Tour de France* in 1948. In 1950 a second black and white channel was created, but it broadcast for only two hours a day. From 1967, its programmes (twelve hours a day) were in colour. It took more than twenty years for the third channel to be made available on French television (31/12/1972). Advertising had been allowed since 1968 on the first two channels, although the third channel was not allowed by the HACA (*Haute Autorité de la Communication et de l'Audiovisuel*) to introduce advertising until later.

In the last twelve years, the PAF (*Paysage Audiovisuel Français*, French Audiovisual Landscape) has undergone several major changes. Half a century after the creation of the first channel, only three terrestrial public channels existed until the Socialists introduced not only the first francophone satellite channel in 1983 (TV5, which brings together French-speaking public television from several countries), but also the first ever encrypted subscription channel in 1984 (Canal+) and two private ones in 1985 ('La Cinq' and 'TV6'). The number of channels therefore doubled between 1985 and 1995.

Planning for the Canal+ project began in 1982. The main characteristic of this channel is its means of support. At the time of its creation, it was decided that this channel would not take revenue from the licence fee, nor from advertising. Its only financing would come from the subscription fees viewers had to pay. Another specific feature was the kind of programmes it was entitled to broadcast. Its conditions of contract (*Cahier des charges*) were radically different from the other existing channels. 40% of broadcasting time was devoted to cinema and 35% to sports and shows. Only 150,000 households had subscribed by the time the first programmes were broadcast in November 1984.

The introduction of the first private channels by a Socialist government was controversial in many ways. For many, it contradicted some of the beliefs advocated by the Socialists as far as audiovisual policies were concerned. The introduction of competition to a French 'PAF' which was already giving cause for concern, and the involvement of Silvio Berlusconi, owner of the commercial channel La

Cinq and whose role in the destruction of Italian public television was a well known fact, were not seen as very positive signs. The usual 'diet' of programmes offered by La Cinq (a mixture of cheap American series, show games and easy entertainment) could also be seen as contradicting the idea of 'culture' advocated by politicians such as the Minister of Culture Jack Lang. More important, however, was the fact that both La Cinq and TV6 were awarded to 'friends' of the French president (see below under HACA). The major change came with the right wing government of Prime Minister Jacques Chirac, which decided to privatise the first channel TF1 in 1987. This privatisation would have major consequences for French broadcasting (see below under CNCL).

Key legislation

This section will describe the different attempts by successive governments to 'liberate' French TV from the power of the state, and will briefly discuss the various organisations created with this aim in mind: ORTF, HACA, CNCL and CSA.

It is often said that French television has undergone more changes than any other institution in France. Before the Socialist Party started a series of radical reforms in the field in 1981, some attempts had been made by the two previous governments to give some form of independence to a medium which was growing fast.

In France, as in many other western European countries, the state has often retained complete control of audio and audiovisual media. For a period from the end of World War II to the victory of the Socialist Party in the presidential elections of 1981, the government in France held the 'reins' of French television, a powerful propaganda instrument that no ruling party was willing to give up. State control was underpinned by the civil service mentality which pervaded French television. De Gaulle is often cited as being the president who, more than anyone else, knew very well how to use such a weapon.

RTF (1945-1964); ORTF (1964-1974)

Television therefore remained under the authority and strict control of the Minister of Information, before moving under its 'supervision' (*tutelle*) from the RTF (*Radio Télévision Française*, a public organization created in 1945) to the 'autonomous' ORTF in 1964. The RTF, established under the de Gaulle government in 1945, represented a state monopoly of audiovisual media. In a way, it might be said that the tough control and domination exercised by the state over French TV 'compensated' for the strong support for the opposition manifested by the French press. Despite its desire to keep the population informed, the state did not modify its monopoly and the situation remained the same even after the RTF become an 'autonomous' organisation.

After the excesses of the state monopoly during the early years of the Fifth Republic, and more especially the private use de Gaulle made of the medium, the RTF was modified and the ORTF (*Office de Radio Télévision Française*) created on June 27 1964 by Alain Peyrefitte, the Minister of Information of the time. According to the law of 27/6/1964, 'the RTF is established as an Office in order to mark in an unquestionable way its specificity as an autonomous public enterprise'. The main aim of this change was to replace the 'authority' of the Ministry of Information with the 'supervision' of not only the Ministry of Information but also the Treasury and Parliament. Thus, in theory, the power of the Ministry was reduced, and was defined through precise texts, rather than through direct intervention as had been the case before. The main failure of the ORTF came from the contradictions between theory and practice. From a financial point of view, despite the relative autonomy of the Office, financial management was under the control of a director general designated by the government and who could be dismissed at any moment.

Another ambiguity in legislation related to programme content. The first article of the law stipulated that: 'The ORTF provides the public service of radio and television in order to satisfy the needs of the audience in terms of culture, education and entertainment'. The fourth article of the law which defined the mission of the ORTF stated that the Board of Directors (*Conseil d'Administration*) had to define the objectives of the Office and had to control *the objectivity and the accuracy of the information broadcast by the Office* (my emphasis). Half of the members of the Board of Directors represented the state and the other half was made up of viewers, journalists and qualified individuals. Here again, the state could intervene in case of disagreement and could reconsider the decisions taken by the Board. Censorship was therefore extremely strong and frequent, and the opposition was denied virtually all access.

This lack of independence of French TV became still more obvious during the so-called 'events' of May 1968. The need for neutral information was such that following the growing discrepancies between the news broadcast by the public media and independent sources (mainly the radio station Europe 1) concerning the events, a majority of journalists belonging to the ORTF went on strike, while others chose to broadcast more objective news. Not only did the government ask for the army to intervene and to occupy some of the transmitting stations, but in addition hundreds of journalists were subsequently sacked. The ORTF lost in this battle what little credibility it had left.

After the issue had been raised during the presidential elections of 1969, attempts were then made to liberalise French TV. In September 1969, the Prime Minister, Jacques Chaban-Delmas, suggested the first changes when he created two autonomous entities and introduced a form of competition between the two existing channels. These two different boards of information were abolished in 1972. In July of the same year a new status for the ORTF was proposed by the government. At the same time advertising was allowed on the first two channels in order to offset the financial problems faced by the ORTF.

Shortly after Valery Giscard-d'Estaing was elected president, the Office was finally dismantled (this was called the *éclatement de l'ORTF*) in 1974 by the Law of August 7 1974. Seven companies were created - three television channels: TF1 (*Télévision Française 1*), Antenne 2 and FR3 (*France Région 3*) - a production company: the SFP (*Société Française de Production*) - a transmission company: the TDF (*Télé Diffusion Française*) - an institute in charge of public radio: Radio-France - an institute in charge of Training, Studies, Archives and Research: the INA (*Institut National de l'Audiovisuel*)

Despite the relative autonomy of the new version of the Office and the fact that the state could not interfere directly, the government was still in charge of appointing the channels' managers. The situation of French TV under Valery Giscard d'Estaing's presidency would remain the same, i.e. on the one hand more autonomy *vis-à-vis* the state, but on the other state control of managerial appointments.

HACA (1981-1986)

The French Socialist president François Mitterrand is often seen as the first to decide to cut the 'umbilical cord' between the state and French television, and the decision was implemented as soon as he was elected (Bonnel, 1989: 329). This idea came up repeatedly during his election campaign. His wish was to create an organisation which would guarantee the complete independence of the television programme makers. The new Audiovisual Bill (29/7/1982), following the recommendations of the Moinot Commission, was published in July 1982. Its first article stated that: '*La communication audiovisuelle est libre*' and the second that: '*les citoyens ont droit à une communication audiovisuelle et pluraliste*'. It introduced the *Haute Autorité de la Communication et de l'Audiovisuel* (HACA) whose aim was to act as an intermediary between the government and the television organisations. The HACA was composed of nine members, of whom three each were chosen by the President of the Senate, the President of the National Assembly, and the President of the Republic, who also nominated the President of the HACA. Their term of office was nine years.

The main aims of the HACA, as stated in the July Bill, were to guarantee the freedom of audiovisual communication and the independence of the public sector, and to monitor the implementation of the *Cahiers des Charges* (conditions of contract) of both the public and private sector.

It had the power to award broadcasting licences, to monitor programmes, to guarantee political pluralism, and to appoint managers. The Bill innovated not only by abolishing the monopoly of previous governments, but also by introducing the possibility of creating private channels (such as La Cinq and TV6) while establishing three new public audiovisual companies: RFI (Radio France Internationale), the *Société de Radiodiffusion et de Télévision pour l'outre Mer* and FMI (*France Média International* which lasted until 1986), and paving the way, as mentioned earlier, for Canal+.

Despite the rather radical changes instigated by the HACA, and the goodwill of its creators, many criticisms can be directed against it. The main problem relates to both its way of financing public channels and its internal organisation. The introduction of competition between the different channels accentuated what was already obvious at the time of the ORTF: the massive importation of foreign programmes, the lack of finance for French audiovisual production, and the increase in running costs. From a political point of view, the way HACA's members were selected (two thirds of the members were nominated by left-wing politicians) was seen as an illustration of the lack of political impartiality. Another example of this impartiality was the award of La Cinq and TV6 to personalities linked to the Socialist Party. Both these points contradicted for some the desire to liberate French television from the state. Additionally, many of the elements proposed in the Bill were at times impossible to implement, since the sanctions available to the HACA were extremely limited.

CNCL (1986-1988)

In May 1986, the Socialist Party was defeated in the general election. France started its first political 'cohabitation' with a conservative Prime Minister and a Socialist President in political opposition to the conservative majority in the National Assembly. The government of Jacques Chirac exploited the failures of the HACA, and soon after his arrival in power, the HACA was dissolved and replaced in accordance with the Bill of September 30 1986 (called '*Loi Léotard*' from the name of the new Minister of Culture and Communication) by the CNCL (*Commission Nationale de la Communication et des Libertés*). The CNCL was composed of thirteen members: two were designated by the President of the Republic, two by the President of the Senate, two by the President of the National Assembly, one by the Council of State, one by the Court of Cassation, one by the National Audit Office, and one by the French Academy. The remaining three members of the CNCL were elected by the other ten. The length of their term of office was the same as before. The desire for political pluralism, which was never really achieved under the HACA, was even less firmly embodied in the CNCL, since only two members were designated by the 'opposition', i.e by the President of the Republic.

The aims and objectives of the CNCL were quite similar to the HACA's as far as the freedom of communication was concerned. It had similar powers: granting licences, monitoring, appointing. It was given more powers to sanction the private sector, as it could suspend or withdraw the licence of the guilty channel and/or even go to court if necessary. Despite these reinforced powers which the previous organization had lacked, the CNCL never had any credibility nor any independence from the political establishment. A witch hunt started soon after the beginning of the first cohabitation. All the managers of public TV or radio channels were sacked before the end of their term and replaced by candidates close to the Gaullist party in power, the RPR or *Rassemblement Pour la République*. All the licences granted by the HACA for *radios libres* (independent local radio stations) were cancelled. Even before the Bill was passed in September 1986, the Chirac government had asked the Council of State to nullify the concessions of La Cinq and TV6, although both channels had just started to broadcast programmes, La Cinq at the end of February and TV6 on 1st March 1986. Both channels were given free to 'buyers' whose political affinities were acceptable to the CNCL.

More important for the future of the PAF than these examples of political impartiality was the desire for liberalism expressed by the government, which wanted to see broadcasting disentangled from the state. Through the voice of its Minister of Culture and Communication, François Léotard, the state announced the era of the '*mieux-disant culturel*'. The main idea behind this rather obscure (and untranslatable) expression, was to favour free enterprise and to introduce free market conditions within the audiovisual field. It is from this perspective that the first and most popular French channel (TF1) was put on sale in 1986 for three billion French francs (nearly half a billion pounds sterling). This was a first in Europe. TF1 was bought by Francis Bouygues, general manager of the world's leading construction company. For some of its toughest critics, in the words of Jean-Jack Queyranne, spokesman for the Socialist Party in May 1986, this privatisation 'was tantamount to the *Comédie Française* in the hands of Coca-Cola or Versailles in the hands of IBM'.

The consequences of the privatisation of TF1 are still visible nowadays as it has radically modified the PAF. First of all, the balance between the private and the public sector became almost the exact reverse of what it had been prior to 1986, with a PAF henceforth for the most part private and only two public channels (Antenne 2 and FR3) out of six. Furthermore, the competition which the government was so eager to create amongst the channels triggered a massive rise in advertising costs (up to 60% between 1986 and 1988). The excessively high price paid by the buyer of TF1 meant a very noticeable increase in cheap imported programmes, mainly from the United States and Japan. The production and creation of French programmes decreased correspondingly, sacrificed for financial reasons. The new *Cahier des charges* of TF1 after its privatisation was much less restrictive in this regard. Thus, some of the former duties of the channel disappeared, among them the obligation to invest 12% to 13% of the budget in the production of fiction and animation, or to spend six million Francs on the broadcasting of French cartoons. Other duties were modified too, such as the compulsory quotas for French language fiction and documentary works, which fell from 450 hours a year for the public TF1 to 250 for the private one.

The CNCL was discredited a few months after its creation following several cases of bribery. François Mitterrand summed it up when he declared in September 1987 that 'the CNCL had not so far achieved anything which could inspire respect'.

CSA (1989-)

At the presidential elections of 1988, the Socialist Party regained power. The new government hoped to avoid previous mistakes (and more especially the lack of political independence which was obvious in the attribution of channels) and instead of completely changing the 1986 Bill, a decision was taken to retain certain aspects of it. The CNCL was, however, replaced by the CSA (*Conseil Supérieur de l'Audiovisuel*) in legislation passed in January 1989. Nine members constituted the new group: three designated by the President of the Republic, three by the President of the Senate and another three by the President of the National Assembly. They had a six year term, which was not compatible with any other professional activities. The first president of the CSA (chosen by François Mitterrand) was Jacques Boutet, former managing director of TF1 between 1981 and 1982.

The tasks of the CSA were quite similar to those of its predecessors. Most of the articles of the *Loi Léotard* were retained, but in most cases the field of action was widened, and some regulations were tightened, especially those regarding the quotas of Francophone and European programmes (see below). The main differences rested in the power to punish channels which did not meet their contractual obligations. As well as suspending the offending channel, the CSA could impose fines.

The main aim of the CSA was to try to save the public sector which had lost its audience in the battle with the private channels (called at the time, '*la guerre de l'audimat*', the ratings' war) and whose situation was critical. Between 1987 and 1989, both Antenne 2 and FR3 had lost almost 30% of market share. In August 1989, a single Director-General for Antenne 2 and FR3, the first '*super-pédégé*', was elected by the CSA, with the power to appoint the managing directors of both channels.

Many decisions had to be taken by the experts of the CSA, called the 'sages' or 'wise ones' since they were chosen for their knowledge of the field and were supposed to act as impartial referees. The survival of the public sector was seen as a priority and several initiatives were launched. The financial situation of Antenne 2 and FR3 was catastrophic. The annual deficit of between one and two billion francs, and the continuously shrinking audience, parallelled by a fall in advertising contracts, was a major concern not only for the CSA but also for the government. The future and the raison d'être of a public sector was questioned. As H. Michel put it: 'What balance could the public sector achieve between its educational objectives and the need to satisfy all those who pay their TV licence? [...] what is their vocation: to win viewers or to satisfy minorities?' (Michel, 1992: 38; editors' translation)

The future of La SEPT (see below) was another issue addressed by the CSA.

Channels

As mentioned above, the PAF has encountered massive changes in the last sixty years. The increase in terrestrial, cable and satellite channels has transformed the audiovisual landscape to such an extent that it is sometimes difficult to list the available networks. If one considers for example the growing expansion of local terrestrial television stations and the belated development of cable with its thematic channels in France, one is a world away from the handful of stations available at the beginning of the 1980s. The list in Figure 1 is even longer if one includes the channels sourced in other countries. To these internal channels should be added RFO (*Radio-Télévision Française d'Outre-mer*, 1982) and TV5 Europe (Francophone channel, 1983).

Most of the channels have already been described. Before turning to the most recent private and public channels, the case of France 3 deserves further attention. Initially created in 1972 as the third public channel devoted to the regions, its status changed after the dismantling of the ORTF in 1974, and even more with the desire for decentralisation of Socialist governments from 1982 on. F3 (then called FR3) suffered 'an identity problem'. Should it compete with or complement F2 (formerly A2)? Could the two public channels (especially since 1989, when the first joint managing director was appointed) both 'cooperate [and] keep their separate identities?' (Pedley, 1993: 167-9).

The principal emphasis will be on the newest channels and more especially Canal Plus and la SEPT-ARTE-La Cinquième which fill the fifth terrestrial network left vacant by the former commercial channel La Cinq. It is in a sense quite ironical that 'the channel of learning and knowledge' (*'chaîne du savoir et de la connaissance'*) as la Cinquième has been dubbed should take up space previously devoted to light entertainment of Anglo-Saxon origin. The collapse in April

ure 1. Table of principal television channels

:hannel name	Date of creation/ privatisation	Ownership and funding	Viewers	Profile
F1	1935/1987	private: Bouygues owner + shareholders advertising	41%	News, light entertainment foreign series
2	1964	State, licence fee advertising (as Antenne 2)	24%	News, fiction entertainment
3	1972 (as FR3;	State, licence fee advertising 1982 for regional documentary networks)	14%	News, fiction
a Cinquième	1994	State	NA	Educational progammes
rte	1993 (La SEPT 1987-1991)	La SEPT (50%)		Culture Franco-German documentaries fictions
anal Plus ncrypted)	1984	Subscriptions Havas, CGE advertising	6%	Sport, films documentaries news
6	1987	CLT, Lyonnaise des eaux + banks advertising	12%	music, fiction entertaiment

1992 of the fifth channel, which previously attracted up to 13% of the French audience, was mainly due to the escalating cost of advertising following the privatisation of TF1. Although La Cinq initially benefited from the liberalisation of the audiovisual domain, it could not compete against its main rival TF1 since they offered similar kinds of programmes. The several attempts by Hachette and Berlusconi to rescue the channel after the media tycoon Hersant (who previously owned the channel) gave up in autumn 1990, failed as losses rose and the cumulative deficit finally reached three billion French francs (Kuhn, 1995: 200). 'La Cinq [...] became the main victim of the liberal, market-oriented television system established in the late 1980s' (Kuhn, 1995: 201).

La SEPT-ARTE-La Cinquième

The cultural channel La SEPT ('SEVEN'), an acronym for *Société d'Edition et de Programmes de Télévision*, was originally planned by the outgoing Socialist government of Laurent Fabius in 1986. Although intended to be broadcast on both satellite and cable, initially, from 8 May 1987, the programmes were broadcast a few days a week on FR3. Joint development with the Germans followed the Mitterrand-Kohl agreement at the 52nd Franco-German summit in December 1988. From September 1989, La SEPT's programmes were shown every Saturday on FR3. Like the other two public channels, La SEPT was a limited company whose capital was divided between the state (25%), Radio-France (15%), l'INA (15%) and France 3 (45%). Since 10 March 1993, La SEPT has owned 50% of the capital of the Franco-German channel ARTE and the company is now called La SEPT-ARTE. The other half of ARTE is owned by Arte Deutschland TV (which is itself controlled by two public German channels). ARTE, the first and only Franco-German channel, was given the frequency vacated by La Cinq after its collapse at the beginning of 1992.

ARTE was created with the aim of promoting the creation, realisation and broadcasting of TV programmes of a cultural nature in order to promote understanding and closeness between European countries. It is broadcast in France from 17.00 on cable and from 19.00 until 1.00 am on the terrestrial channel. Since March 1993, it has also been available on Belgian cable.

The education channel La Cinquième (or TSF: *Télévision-Savoir-Formation* as it was first called) was a project initially launched by the Socialist Minister of Culture, Jack Lang, at the beginning of the 1990s and therefore before the political change of 1993, when a second cohabitation started. It was then planned that it would fill the gap on the fifth channel before the start of ARTE broadcasts. The Carignon Bill (from the name of the then Minister of Communication of the right wing Balladur government) passed on February 1 1994 acknowledged the necessity of such a channel. The idea of an educational channel generated some hostility, as many feared that television was encroaching on territory previously the exclusive domain of the Ministry of Education. However, the presence within the comittee of specialists in education, teachers and researchers gave credibility to the project.

The main innovation was the co-existence of two TV channels on the same frequency. Another was the fact that numerous ministers were involved in the project, including the Ministers of Education, of Youth and Sports, and of Employment. The task of the channel was first of all to offer programmes of a social, cultural and educational nature. In addition to its educational programmes, it aimed to give information about work and training. Existing documentaries were used together with programmes made specifically for the channel. Special attention was given to the necessity of harmonising the content of the broadcasts with those of ARTE. The budget of the channel is between 600 million and 1 billion French francs. The first daytime programmes started in December 1994.

Canal Plus

This channel is another innovation in the PAF. As mentioned earlier, it was created in 1984 and was the first ever encrypted subscription channel. Canal Plus is financed mainly through subscriptions (83.5%). The rest of its budget comes from advertising and sponsorship (4.5%), and miscellaneous (11.6%). Its role in financing, producing and protecting the French cinema industry is enormous, as it is obliged to invest 25% of its profits in acquiring broadcasting rights to films (60% of European origin, of which two-thirds, or 40% of the total, are Francophone). Canal Plus is also the French channel which invests the most in '*préachat*' (advance purchase).

Its success (although it was not immediate) can be partly explained by the lack of competition, as cable television was very slow to start in France in the 1980s. The mixture of very popular programmes (films, sports and documentaries) it offers and its unique profile are other important factors to consider. From only 2 million subscribers in 1987, the channel reached 3.7 million subscribers in 1994, having reached a plateau in subscriptions in 1993, for the first time since its creation. Its success, however, has allowed the channel to spread within Europe and beyond. Canal Plus started to broadcast in Belgium in 1989 (150,000 subscribers in 1994) and in Spain in 1990 (890,000 subscribers in 1994) and Germany (810,000 subscribers in 1994). It is investing in Francophone Africa as well (45,000 subscribers in 1994).

The channel has recently encountered serious problems which have altered its image of 'successful outsider': not only has it lost more money than it has earned in 1994 and gained no new subscriptions between the end of 1993 and September 1994, but it has also seen its share falling in 1994 to the 1990 level. Various explanations may be advanced, including the competition of cable and thematic channels, and the sudden resignation of its creator André Rousselet in February 1993, following politico-economic imbroglios.

The recent decision by the managing director Pierre Lescure to sign a contract with the German group Bertelsmann (21/07/1994) should allow the channel to regain its success through widespread expansion of its services across Europe.

Penetration

As mentioned before, France was very late in developing its audiovisual network. Therefore the acquisition of TV sets in French households was very slow to start. In the 1960s, which is often seen as the starting point of French television, or at least the moment when France developed to a fuller extent its television network (with the creation of the second channel in 1964 and the beginning of colour television in 1967), the number of television sets per 1000 inhabitants was 41 for France, which is a low figure compared with other countries (such as the United States with 310, or Great Britain 211 at the same time), but high compared with 1955 when the figure was only three television sets per 1000 inhabitants. In terms of percentages of households with television, the figures are 6% in 1958 and 25% in 1963. At the end of the sixties, 66.4% of French households were equipped with television. Between 1980 and 1992, the percentage grew from 90% to 96.6%, with the proportion of households owning more than one television set rising from 23% in 1988 to 33.5% in 1992, and the proportion of households with only one set decreasing by almost 10% during the same period.

Funding

The system of funding in France varies considerably from one channel to another, both in the public and the private sector. As far as the former is concerned, the main resource is the 700 FF licence fee. Advertising (allowed with some restrictions) represents about 25% of the budget and has not noticeably increased since 1987. Since 1988, the state has also contributed to the financing of the public sector, while commercial profits obtained from '*produits dérivés*' (spin-offs) have fallen.

The licence fee represented 34.1% of public sector income in 1986. In 1992, it was only 21%. At the same time, it financed 74% of the budget of F3 and 52.8% of F2 in 1992 (as against 74% and 25.7% respectively in 1987). The licence fee paid in France is one of the cheapest in Europe (Cortade, 1993: 40-41).

The growth of the private sector over the last ten years has meant increasing development of advertising as a major source of profits for the commercialised channels. The introduction of competition within the PAF (see above, section 2, CNCL) triggered a 'dirty war' between the public and the private sector as the cost of advertising escalated after the privatisation of TF1 in 1987.

Programme production

French channels, both private and public, have to obey certain rules regarding their programmes. One of the CSA's roles, as mentioned above, is to prepare a set of regulations (*cahier des charges*) for each channel. The different Bills which have been passed in the last 15 years were aimed at giving more and more power to the regulatory bodies (the HACA, the CNCL and the CSA) as far as the channels' duties were concerned. The Bill of January 17 1989, which introduced the CSA to replace the CNCL, also brought in several modifications to the law of September 30 1986. The most important element was the idea of implementing rules regarding the creation and broadcasting of '*oeuvres audiovisuelles d'expression française*'. This point was nonetheless present in the 30/09/86 Bill (article 45): 'Programming is designed to promote knowledge and defence of the French language, and to illustrate the expression of francophonie worldwide'. This aspect of the law was reinforced by the orders of January 17 1990, whose aim was to help the production of French and European programmes in order to counteract the massive imports of products from non-European countries. The first of these orders (17/1/1990

Funding of the public sector in France (1987-1993) in million FF (the figure in bracket indicates the total budget for the year)

	State	Licence fees	Advertising	Produits dérivés (spinoffs)
1987 (9,516)	0	4,000	2,126	3,388
1988 (10,209)	180	4,440	2,217	3,372
1989 (10,711)	236	4,541	2,358	3,576
1990 (12,025)	250	5,238	2,351	4,185
1991 (8,846)	575	5,612	2,062	596
1992 (10,135)	845	6,509	2,257	523
1993 (11,330)	1452	6,520	2,775	582

but implemented only from April 1 1992) introduced quotas in order to promote French and European programmes. Channels were henceforth obliged to broadcast 50% French-language programmes and 60% programmes of European origin during prime-time (i.e. from 14.00 to 23.00 on Wednesdays and from 18.00 to 23.00 the rest of the week).

As far as the production of French and European programmes is concerned, the January 1990 order offered two possibilities to the channels: either to invest 15% of their annual turnover in commissioning French programmes and to broadcast 120 hours a year between 20.00 and 21.00; or to invest 20% of their annual turnover in European programmes, 15% of which must be in French.

The main problem encountered by French channels (both public and private), is the lack of French-sourced production. Only 5% of total broadcast time is filled by home-made products. This applies both to fiction and to documentaries. The reason for this critical situation is on the one hand the very severe international competition, and on the other (but both are linked) the financing of domestic production. In 1991, the hourly cost of producing French drama was 3.5 MF, while the hourly cost of rights to American television drama varies between 0.3 and 2.5 MF (Cortade, 1993: 65). The privatisation of TF1 and the introduction of competition between the channels had weakened the previous role of the SFP. Production companies, numbering 560 in 1990, vary in type and size. On average, around 50 new companies have been created every year since 1982. The majority were created at the same time as the commercial channels. Since the deregulation of the market, many newcomers (banks, media groups) have started to invest in the production of audiovisual programmes. Another source of financing has emerged from the creation of the SOFICA (*Sociétés de Financement du Cinéma et de l'Audiovisuel*) in 1985.

Satellite and cable

Compared to other European countries (and especially to Germany), France initially failed to develop cable television during the 1980s. In 1982, a 'Plan Câble' was launched by the newly elected Socialist government led by Pierre Mauroy. It was planned that '1.5 million households would be linked to a cable system by 1986, with 1 million additional households being connected every year after that' (Kuhn 1995: 212). The state-owned company France Telecom was in charge of the 'cabling' of France. The main private investors, whose task was to commercialise cable television, were La Lyonnaise des Eaux-Dumez, La Générale des Eaux and La Caisse des Dépôts.

According to the national survey organisation INSEE, only 428,500 households were connected in 1987, of which only 58,000 subscribed. The deficit reached 3.6 MF in 1991. Different reasons are suggested: not only was there an obvious lack of coordination between the different parties involved, but the subscription fees were considered by the general public to be too high. There were complaints as well regarding the content and the quality of the programmes offered. The competition of Canal Plus is also an important factor to consider. The lack of interest in the new medium meant a lack of resources, especially when between 20% and 30% of the subscribers decided to cancel their subscriptions.

However, after 1991 when only 13% of connected households actually subscribed (the profitability threshold was 45% take-up), the situation seemed to improve. Several attempts were made not only to improve the quality of the programmes but to reduce the subscription. In 1992 the figures reached a more positive 4,644,899 connections and 1,032,452 subscriptions. In November 1993 there were more than 5,000,000 connected and 1,228,000 subscribers (i.e. about 25%). These figures are growing steadily and it is now an accepted fact that France is finally succeeding in developing its cable network.

Links with cinema

French cinema has always been heavily subsidised by the state. Since 1948, several attempts have been made to provide financial support to the French cinema industry, but until the 1980s, French television was not directly involved in this process. Since the creation in 1983 of the IFIC (*Institut pour le Financement du Cinéma et des Industries Culturelles*), French channels have been asked to help directly or indirectly to support French cinema. Thereafter, succesive legislation extended the levy. A new tax, introduced in 1986, represented 3% of annual turnover; it was raised to 4% in 1987 and 5.5% in 1992.

French channels also became very much involved in the production of films during the second half of the 1980s, and it can be said that almost all films produced in France were partly financed by French television. Its contribution to the production of French films grew from a little over 9% in 1985 to a little over 30% in 1992. As far as the broadcasting of films on French channels is concerned, several rules have been introduced by the CSA in order to protect the French cinema industry. Thus, for example, channels have to wait for three years (only two if they are co-producers) before broadcasting a film. They are not allowed to broadcast films during the weekend or on Wednesday or Friday night. The channels must respect the same quotas as for other programmes (i.e. 60% of films of European origin, and 40% of French-speaking films). Canal Plus follows a different pattern and does not have to meet the same obligations, since it is entitled to show a higher proportion of films as long as it reinvests the profits in film production. The different restrictions regarding broadcasting schedules and the length of time which must elapse between the public release of a film and its broadcasting do not apply to this channel.

Conclusion

Despite all the reforms of the last fifteen years, French television is currently in very bad shape. Freed from overt political influence, it now seems to be struggling against the influence of money. The opening declaration at the European Audiovisual Conference in the summer of 1994, by Jacques Delors, then President of the European Commission, that 'audiovisual material is not a piece of merchandise like any other' might sound rather optimistic when one considers what is happening to the French public and private channels.

Key dates

1935	First programmes
1945	Creation of the RTF after the war
1950	Second channel starts
1964	Creation of the ORTF
1968-1972	Introduction of advertising
31-12-1972	Creation of the third channel
1974	Dismantling (éclatement) of the ORTF into seven companies
1982	Creation of the HACA
	Launch of Canal Plus

1985	Creation of two private channels: La Cinq and TV6
1986	New bill on broadcasting: creation of the CNCL
	Privatisation of TF1
	La Cinq and TV6 are reattributed to new groups
1989	Creation of the CSA
1990	Planning of ARTE
1992-1993	Collapse of La Cinq (April 1992)
	ARTE starts broadcasting on the fifth channel (after 7 pm)
1994-1995	Planning of an educational channel (La Cinquième) on the fifth channel (before 7 pm). Programmes start in December 1994.

Abbreviations

ARTE:	Association Relative à la Télévision Européenne
CNCL:	Commission Nationale de la Communication et les Libertés
CSA:	Conseil Supérieur de l'Audiovisuel
F2:	2nd channel
F3:	3rd channel
HACA:	Haute Autorité de la Communication et de l'Audiovisuel
INA:	Institut National de l'Audiovisuel
M6:	6th channel
ORTF:	Office de Radio Télévision Française
PAF:	Paysage Audiovisuel Français
PTT:	Postes, Téléphones et Télégraphes
RTF:	Radio Télévision Française
SEPT:	Société d'Edition et de Programmes de Télévision
TDF:	Télé Diffusion Française
TF1:	1st channel

Bibliography

Astier, I. (1993), 'Le Mystère de la 5', *Médiaspouvoirs*, Paris, no 30.

Bahu-Leyser, D. (1993), 'Histoire des équipements et des pratiques télévisuelles en France', *Médiaspouvoirs*, Paris, nos 31-32.

Bonnel, R. (1989), *La vingt-cinquième image: une économie de l'audiovisuel*, Paris, Gallimard.

Cayrol, R. (1991), *Les médias*, Paris, PUF.

Chapman, R. and Hewitt, N. (eds.) (1992), *Popular Culture and Mass Communication in Twentieth Century France*, Lampeter, The Edwin Mellen Press.

Les Chiffres-clés de la télévision française (1990), Paris, INA, CSA, 1990.

Cortade, J.-E. (1993), *La Télévision française 1986-1992*, Paris, PUF.

Dargnies, S. (1989), *Les Chiffres-clés de la télévision française (1988-1989)*, Paris, INA, CSA.

Delivet, J.-P. (1989), 'Une nouvelle instance de régulation: le Conseil Supérieur de l'Audiovisuel', *Médiaspouvoirs*, Paris, no 14.

Forbes, J. (1995), 'The Rise of Audiovisual Culture', in J. Forbes and M. Kelly (eds), *French Cultural Studies*, Oxford University Press.

Hewitt, N., and Rigby, B. (eds.)(1991), *France and the Mass Media*, Macmillan.

Kuhn, R. (1995), *The Media in France*, Routledge.

Le Diberder, A. (1988), *Briser les chaînes: une introduction à l'après télévision*, Paris, La découverte.

Lunven, R., and Vedel, T. (1993), *La télévision de demain*, Paris, Armand Colin.

Le Monde dossiers et documents (1988), *La télévision en 1987: Le grand chambardement*.

Mamère, N. (1988), *La Dictature de l'audimat*, Paris, La découverte, 1988.

Pedley, A. (1993), 'The Media' in M. Cook (ed.), *French Culture since 1945*, Longman.

Rigby, B. (1991), *Popular Culture in Modern France*, Routledge.

Television in Germany

John Sandford

History

As with many other media, there are competing claims about which country was the 'inventor' of television. At least there is no dispute that Germany was well in the vanguard, both in technological developments from the latter part of the nineteenth century onwards, and in the introduction of a regular programme service in the mid-1930s. Nazi ambitions to turn the fledgling medium into an instrument of mass propaganda were thwarted by the outbreak of the Second World War, and the decision to concentrate limited resources on the manufacture of radio sets. Nonetheless, television transmissions to the country's 500 or so receivers (most of them set up in public places) continued right through to 1944, whilst in occupied Paris a Franco-German service broadcast for between five and eight hours a day from the Eiffel Tower for some twenty months from the beginning of 1943.

After the defeat of Nazi Germany, the Allies re-established broadcasting on the basis of separate services in each of their Zones of Occupation. Initially only radio services were set up, with organisational and financing structures based on the BBC's 'public service' principle, not least because the Allies were determined to avoid the kind of state control that had existed in the Nazi years, and because the other option - an American-style commercial system - was inconceivable in the economic circumstances of the immediate post-war years. The '*Landesrundfunkanstalten*' - or 'Land broadcasting corporations' - that came into being under Allied supervision set the pattern for public broadcasting in Germany that remains operative right through to the present day.

The distinctive regional structure of German broadcasting - partly a resurrection of the shape of radio in the Weimar Republic, partly a result of the zonal division of post-war Germany, and partly a further reflection of the desire to create a system that would be less prone to centralised state control - meant that the rebirth of television occurred separately in different parts of the country. The largest, wealthiest, and most enterprising of the new broadcasting corporations - the *Nordwestdeutscher Rundfunk* (NWDR), which served the former British Zone of Occupation - transmitted the first post-war German television picture on 12 July 1950, and followed this with a regular experimental service on alternate days just over four months later, with daily transmissions beginning on Christmas Day 1952. Over the course of the early fifties other *Landesrundfunkanstalten* began television transmissions, but initially the arrangements for programme sharing were ad hoc and haphazard. It was clear that the complexity and expense of the new medium made a degree of coordination and pooling much more necessary than was the case with radio, and this factor was a major impetus towards the establishment in November 1953 of the '*Deutsches Fernsehen*', or 'first channel', as it was to become known.

The coordination of the *Deutsches Fernsehen* is the responsibility of the 'ARD', the mercifully brief abbreviation for the cumbersomely-named '*Arbeitsgemeinschaft der öffentlich-rechtlichen Rundfunksanstalten der Bundesrepublik Deutschland*', which had been established in 1950 to represent the joint interests of the several Land broadcasting corporations. (In fact, the common practice in Germany is to refer to the first channel itself as 'the ARD', even though strictly speaking this organisation's remit goes much wider.) The pattern established in 1953 persists to this day, with peak-time and daytime programmes networked nationally, and the early evening reserved for regional items. As with Britain's ITV network, where a kindred pattern obtains, the proportion of programmes contributed by the various regions is determined by a formula worked out on the basis of their size and resources.

Moves by the Adenauer Government in the late fifties and early sixties to introduce a second television channel were met with resistance not only by the ARD corporations, but also by a number of Land governments, who successfully appealed to the Federal Constitutional Court both for an injunction restraining the Federal Government from starting the new service and for an adjudication on its legality. On 28 February 1961 the Court delivered its verdict, declaring that the Government's proposal was indeed unacceptable on two main grounds: firstly because it breached the principle that broadcasting was a matter for the Länder and not the Federation, and secondly because a privately-owned service (which was what the proposal entailed) could not offer the same guarantees of freedom of expression as did the pluralist structures of the existing public service corporations, especially given the limited number of channels that were - at that time - technologically feasible.

The 1961 'Television Verdict' was to turn out to be a milestone in German broadcasting history, ensuring in particular that it would be more than two decades before the Federal Republic belatedly joined in the Europe-wide trend towards the introduction of commercial radio and television. The Länder, taking their cue from the Court, lost no time in setting up a new public-service programme with the name '*Zweites Deutsches Fernsehen*', which began broadcasting in April 1963. Although the Mainz-based ZDF provides a single country-wide service, the fact that it is based on a state treaty between the Länder rather than on federal law ensures its compliance with the constitutional requirements elaborated in the 'Television Verdict'. The following year viewers in Bavaria and Hessen were provided with third channels by their regional corporations, and by the end of the decade regional third channels (in the southwest and north of the country organised as joint operations of adjacent corporations) were available throughout the Federal Republic.

This pattern - of two main channels (ARD and ZDF), and a third regional one - was to be the norm for West German viewers throughout the 1970s and for the first half of the 1980s. Pressure from interested parties for the introduction of commercial provision was met with resistance from the existing public corporations and such groupings as the churches, unions, and - not least - the SPD-led government of Helmut Schmidt. The eventual establishment of commercial TV was facilitated by technological advances (cable and satellite), which met some of the reservations of the 1961 Constitutional Court verdict, and, above all, by the change of government in Bonn that occurred in the autumn of 1982. Constitutionally barred from acting directly in the sphere of broadcasting, the new Kohl government used its control of the Federal Post Office to initiate a massive countrywide programme of cable-laying that not only gave the Federal Republic in a very short time the biggest cable network in Europe, but also - as the government had intended - destroyed the argument of the opponents of commercialisation that there were not enough channels available. At the same time, the coming of satellite television confronted SPD-led Länder with services that offered no respect to whatever laws that might apply within their own geographical boundaries. And so, in 1984 commercial television finally arrived in Germany in the shape of RTL-Plus and SAT 1, the two channels that have remained the market leaders despite the establishment since then of numerous competitors.

The 'dual order' of public and commercial provision was well established by the end of the decade, and the coming of German unification in 1990 altered the map, but not the basic principles, of the Federal Republic's broadcasting system. In fact, East Germans had long been regular viewers of Western television: programmes could be received throughout much of the GDR, and - in the area of news and current affairs at least - provided an influential alternative to the explicitly propagandistic offerings of the two centralised, state-run channels of GDR television. Contrary to the wishes of those many East Germans who had hoped after Unification to preserve at least one public channel that would reflect their peculiar historical experience and identity, the West German pattern was adopted wholesale, with two new corporations (*Mitteldeutscher Rundfunk* and *Ostdeutscher Rundfunk Brandenburg*) being set up to serve the south and centre of the former GDR, while East Berlin and the northern *Land of Mecklenburg-Vorpommern* were incorporated into the already existing *Sender Freies Berlin* and *Norddeutscher Rundfunk* respectively.

Key legislation

Because of the federal structure of the country, broadcasting law in Germany is highly complex, consisting of an interlocking set of Land, inter-Land, and - more recently - European enactments. As all legislation must accord with the Constitution, the Federal Constitutional Court has played a crucial role in setting the parameters for the changing media landscape.

The Court's starting point has been Article 5 of the Constitution, which guarantees freedom of expression, as well as the freedom of the press and broadcasting. The already-mentioned 'Television Verdict' of 1961 was to turn out to be the first of a series, in which the original principles have been refined and developed in the light of changing circumstances (Humphreys, 1994: 338-42). All have turned in one way or another on the issue of pluralism and the assumption that a healthy democracy is predicated on the possibility for the full range of opinions to find expression in the media. The ruling that cleared the way for commercial broadcasting came in 1981, when the Court introduced the terms 'internal pluralism' and 'external pluralism' into the debate: the former was typified in the structure of the existing broadcasting corporations, where pluralism was ensured within the individual organisation, while 'external pluralism' characterised the press, where the range of opinions on offer existed between the different competing publications. This judgment, technically the 'Third Television Verdict', was seen as giving the green light for a commercial television system, provided that - like the press - it was characterised by a multiplicity of outlets.

In 1986, in its 'Fourth Broadcasting Verdict', as these judgements were now to be known, the Court, whilst recognising that a 'dual order' had come into being in German broadcasting, made the important point that the public corporations were still entrusted with that 'essential basic provision' of breadth, variety, and quality in programming that could not reasonably be expected of commercial services. Indeed, this ruling in effect made the continuing operation of public broadcasting a precondition of the admissibility of commercial services, and its practical implications were further clarified in the 'Fifth Broadcasting Verdict' of 1987, in which the Court stated that the public corporations enjoy a 'guarantee of continued existence and development' - meaning in particular that they may not be put at a competitive disadvantage through inadequate funding, or exclusion from technical advances or the provision of new types of service such as local broadcasting.

Regulatory frameworks

Broadcasting in Germany, whether it be public or commercial, is subject to the normal laws of the land and the provisions of the Constitution, but its operations are more immediately determined by specific laws of the individual Länder. Before the advent of commercial programmes, this meant simply the 'Broadcasting Laws' that spell out the role and structure of each of the public broadcasting corporations. (In the case of the ZDF and those corporations that are shared by more than one Land, inter-Land 'State Treaties' fulfil the same function.) The general principles of these laws are the same, and provide for the democratic and pluralist accountability of each corporation through the institution of the 'Broadcasting Council': a governing body whose members represent a range of interests (technically designated as 'socially significant groups') from within the region concerned as well as - on a proportional basis - the political parties of the relevant Land parliaments. More day-to-day decisions are taken by the 'Administrative Council', the next level in the administrative pyramid, at the apex of which is the 'Intendant', the Director General who, like the Administrative Council, is ultimately appointed by, and accountable to, the Broadcasting Council. All levels of the corporations are bound by the general principles spelt out in the Broadcasting Laws, which typically prescribe respect for, and promotion of, the democratic order, and the provision of programmes that offer information, education, and entertainment.

The coming of commercial radio and television has meant that these original laws on public broadcasting have had to be supplemented in each Land by a second law to regulate the operations of the new private sector. These 'Land Media Laws' contain provisions for the licensing of commercial operators, and for their supervision by 'Land Media Authorities'. These latter bodies are the equivalent of the public services' 'Broadcasting Councils', and similarly include representatives of 'socially significant groups'. Initially regarded with scepticism by critics of commercial broadcasting, these Media Authorities have latterly begun to constrain the private services to adhere more closely to the requirements of their licences. The provisions of the individual Länder are given an overall framework in the shape of the lengthy 'Inter-Land Treaty on Broadcasting in the United Germany', which came into force at the beginning of 1992, replacing an earlier West German treaty of 1987.

Links with government

The German federal system leaves matters of '*Kultur*' to the jurisdiction of the individual Länder, and, as this heading includes broadcasting, the country's radio and television services fall outside the competence of the federal government and parliament in Bonn. This principle was tested and confirmed by the Federal Constitutional Court in its 1961 'Television Verdict' and has always been jealously guarded by the Länder themselves. The Court's further affirmation of the principle of '*Staatsferne*', which implies an 'arm's-length' relationship between broadcasting and the state (at any of its levels), indicates that the existing mix of independent public corporations and commercial services may not be supplanted or joined by any form of state-controlled broadcasting.

This does not, of course, mean that German radio and television are free of political influences and constraints. Land governments in particular can be given to attempting to exert considerable pressure on 'their' regional corporations, although in theory their power to do so is limited by the proportional mix in the make-up of the Broadcasting Councils. The setting of the licence fee - a joint task of the Land governments - has also long been seen as an area in which broadcasting corporations were open to intimidation: however, the contrasting political constellations of the Länder have traditionally diluted this danger, and in any case, as a result of a verdict of the Federal Constitutional Court of February 1994, which criticised the undue political interference to which the present system could lead, licence fees are in future to be determined by an independent commission.

The 'socially significant groups' - such as churches, educationalists, unions, or employers' organisations - that provide the majority of members of the Broadcasting Councils themselves often manifest a distinct political hue, and appointments within broadcasting corporations tend to be much more visibly political than is usually the case in, for instance, Britain, with party membership often playing an important role in the filling of senior positions. The result is that some corporations are traditionally perceived as being on balance more 'red' (e.g. the Westdeutscher Rundfunk) or 'black' (e.g. the Bayerischer Rundfunk) than others. However, total political uniformity is avoided by the distinctive German practice of *'Proporz'*, which ensures that posts are also allocated to members of opposition parties (though this is a matter of tradition rather than law). More than anything, though, it is the system's federal structure that means that overall the networked service maintains a broad balance in political matters.

The 'lighter touch' that applies in the commercial sphere, together with the application here of the principle of 'external pluralism' - in other words, a regime that is closer to that of the press - has led to concerns that the political balance enjoined upon the public corporations cannot be guaranteed in the new media. Indeed, this was a major issue in the debate that preceded the introduction of commercial broadcasting, where sceptics and opponents were to be found broadly speaking on the left of the political spectrum. In the event, commercial television in Germany has turned out thus far to be comparatively neutral in its politics, with any political leanings in the two main channels (RTL and SAT 1) divided between the two main parties (SPD and CDU respectively).

Funding and ownership statistics

Public service broadcasting in Germany has always been partly 'commercial' inasmuch as its funding has come from a mix of licence-fee revenue and advertising. Only the regional 'Third Channels', the minority public-service channels 3 SAT and arte, and the pay-TV channel Premiere are completely free of advertisements. For the ARD corporations the licence fee, which is levied on set-owners and currently stands at DM 23.80 a month, remains the major source of income; the ZDF on the other hand, because of its nation-wide service, is able to charge higher rates to advertisers, and the proportion of its revenue from this source is somewhat larger. The ZDF receives 30% of the total revenue from that part of the licence fee that is dedicated specifically to television, with 70% going to the ARD corporations. The volume of advertising permitted on the public service channels is strictly limited: thus it may not exceed twenty minutes in total daily, is limited to the period between 6 p.m. and 8 p.m., and may not interrupt programmes. Growing competition for advertising with the private channels has severely eroded the income of public broadcasting in recent years - thus advertising revenue, which formerly accounted for some 20 percent of the ARD corporations' income, shrank during the first half of the nineties to less than 5 percent - as a result of which the coporations have campaigned for the lifting of some of the restrictions - in particular the 8 p.m. deadline - to which they are subjected (Berg, 1995).

The private channels, which rely almost entirely on advertising for their income, are subject to a much laxer regime, and may devote as much as 20 percent of their airtime to commercials. Funding through subscription, on the other hand, is undeveloped in Germany, with only the movie channel Premiere, with 1.2 million subscribers, operating as a pay-TV service. (Recently introduced commercial digital services are also being funded this way, though initial take-up has - because of consumer resistance to investment in the necessary equipment - been minimal.)

Statistics on television set ownership indicate that in 1995 just over 32 million receivers were registered, representing 396 sets per thousand head of the population, with television available in 32.45 million of the country's 32.81 million households. The proportion of households with cable connection stood in 1995 at 64%, with 41% (=15.5 million households) actually making use of the facility, and 9.2 million households used satellite dishes, just over half of which were communal installations (Media Perspektiven Basisdaten, 1995: 5 - 9). The distinction between 'satellite', 'cable', and 'terrestrial' channels is much less meaningful and clear-cut in Germany than in Britain, as to all intents and purposes all channels are available on cable, the great majority also on the Astra satellite, and all the ARD and ZDF services together with an increasing proportion of the commercial channels via traditional terrestrial transmissions.

Available channels

Because of the predominantly regional structure of public service broadcasting in Germany, the overall number of channels was, by European standards, comparatively high even before the proliferation of commercial services, and, given the 'spill-over' of transmissions from adjacent regions, as well as from the GDR, Austria, and Switzerland, many households could already receive upwards of half a dozen German-language programmes as early as the 1970s. In the 1980s the advent of commercial television, as well as the spread of cable and satellite services, brought a steep increase in the number of available channels, so that in the mid-nineties the average German household has potential access to some forty German-language programmes.

However, viewing figures indicate that for some time four channels in particular have remained the most frequently watched. Two of these are the original German services: the ARD's 'First Channel' and the ZDF. The other two were the first of the commercial channels to be established: RTL and SAT 1. The two main public service channels provide a diet of programmes that is in many ways akin to that of BBC1 in Britain: a mix of information and entertainment, with the day's offerings being clearly punctuated by 'flagship' news bulletins.

RTL and SAT 1, on the other hand, are much more unashamedly populist in their programming and presentation, though they too have developed a number of more serious slots (typically in the late evening) for documentary and current-affairs series.

The 'Third Programmes' of the ARD corporations were originally the preserve of regional and 'highbrow' programming. Most of them are in fact now available nationwide on satellite and/or cable, and, although their regional remit is still important, they have all determinedly shaken off their former 'elitist' feel. One channel, however, has expressly promoted its serious (and 'advertisement-free') image, and that is 3 SAT, a service provided jointly by the public broadcasters of Germany, Austria, and German Switzerland. Even 3 SAT, though, is far outdone in terms of earnest cultural appeal by the programmes of the decidedly minority channel arte.

Alongside the 'big two', RTL and SAT 1, a range of commercial channels exists that can be divided roughly into those offering general entertainment - much of it bought in from America -and those offering specialist programming. The former category consists of Pro7, Kabel 1 (despite its name no longer exclusively a cable service), Vox (originally launched as a 'serious' channel but very quickly 'down-marketed' in pursuit of more satisfactory ratings), and two further services from the RTL stable, RTL 2 and Super RTL. The specialist channels include the two sports channels DSF and Eurosport, the news channel n-tv, the MTV-style teenager channels Viva 1 and Viva 2 , and - the latest addition, launched in August 1995 - the women's channel TM 3. Local channels are also available to cable subscribers in some cities, as are the so-called 'Open Channels', introduced in the early 1980s as experiments in viewer-participation and which provide facilities for members of the public to make their own programmes.

Viewing figures have shown a steady erosion of the dominance of the ARD and the ZDF since the introduction of commercial television, and by the early nineties the point had been reached where the figures for RTL finally overtook those of either of the public services. Statistics for 1995 show the overall share of viewing at 17.6% for RTL, 14.7% each for the ZDF and SAT 1, and 14.6% for the ARD first channel. (The ARD's figures in fact look more favourable if one adds to the national first programme the 9.7% that goes to the regional third programmes.) With the exception of Pro 7 (9.9%) and RTL 2 (4.6%), none of the remaining channels takes more than a 3% share of the audience (Darschin and Frank, 1996: 176).

Such global figures are obviously affected by a range of factors and hide a spectrum of other pertinent statistics. Clearly, for instance, meaningful comparisons of the 'popularity' of programmes cannot be made unless they are all equally available to viewers: as yet this is broadly true only of the ARD, ZDF, RTL, and SAT 1. Viewing preferences also differ across age groups (with children particularly attracted to the commercial programmes), and from one time of the day to another (with RTL and the smaller commercial channels accounting for a disproportionately large share of daytime viewing, followed by a swing to the public service channels in the early evening). There is moreover intriguing evidence of marked differences in the viewing habits of East and West Germans since Unification. Thus, whereas in 1995 television sets in the western Länder were switched on for an average of 289 minutes a day, and watched for 181 minutes by adults and 93 minutes by children, the corresponding figures for East Germany were roughly a fifth higher at 331, 207, and 101 minutes respectively. It is also clear that East Germans give greater preference to the commercial channels than do their Western counterparts (Darschin and Frank, 1996: 176-7).

Conclusions

In some respects, and certainly compared with Britain, Germany was a 'late developer' in the television age: it was not, for instance, until the 1960s that the majority of German households had acquired a television set - something that had happened in Britain almost a decade earlier; and it was not until the eighties that commercial television was introduced into Germany - more than a quarter of a century after Britain, and later than in many other European countries. Today, with their near-saturation ownership of television sets and access to some thirty different domestic channels, the 80 million residents of the Federal Republic have become the leading 'television nation' in Europe.

The transformation has not occurred without comment or criticism. The 1970s in particular saw a lengthy debate waged between the proponents and the opponents of commercial broadcasting and multi-channel television. While those in favour (keenly supported by many who stood to gain financially) spoke of quantitative gains in democracy and pluralism, others saw the issue as one of qualitative loss - of standards, accountability, independence, quality of life, and - a peculiarly German twist to not unfamiliar arguments - of the socially 'integrating' function of a limited number of channels (Humphreys, 1994: 193-238; Sandford, 1985).

In retrospect it seems inevitable that the defenders of the status quo should have lost, and that German television should have embarked on the path - albeit after some delay - already trodden by much of the rest of the industrialised world. Today, the focus of debate has accordingly shifted, and it has become concerned less with fundamental principles than with the daunting intricacies of the emerging media empires, manifested in particular in the almost impossible task of untangling the webs of whole and partial ownership, control, and influence that are accumulating across the spectrum of television, radio, press, and new media both within Germany and the wider world. The legal position regarding cross-ownership is correspondingly complex and fluid, and differs from Land to Land; the effectiveness of the 'Commission for the Ascertainment of Concentration', agreed on by the Länder in October 1995 as a means of advising Land Media Authorities on the licensing of broadcasters whose programmes extend beyond Land boundaries, has yet to prove itself.

For all the byzantine structures that now exist, dominant control of the kind experienced in some other countries has thus far failed to develop in Germany. Three players in particular have emerged as the most-discussed participants in the game of media empire building: Bertelsmann, the second biggest media concern in the world and a major shareholder in RTL; Springer, the biggest of Germany's newspaper and magazine publishers and a major shareholder in SAT 1; and - above all - Leo Kirch, who has risen since Axel Springer's death in 1985 to become the German 'media mogul', with shares in SAT 1, Premiere, and DSF, ambitions in the field of digital TV, a virtual stranglehold on television film rights through his ownership of a vast film library, and influence, to say the least, on Pro 7 and the Kabelkanal via his son Thomas, whom Kirch has described as his 'successor' (on Kirch see Dähnhardt, 1992).

Concern with the structures of commercial television has not diminished recurrent interventions in the area of public service broadcasting - typically by conservative politicians arguing that the ARD in particular is politically biased and/or enjoys privileges that put its corporations at an unfair advantage, while politicians of the left are given to asserting precisely the opposite. Such arguments have to some extent helped shape the new forms of German television, but the German tradition of recourse to the law in matters that elsewhere might be the subject of a purely political decision means that one can be sure that the Federal Constitutional Court will be called upon to deliver yet more 'Broadcasting Verdicts' as the television landscape continues to develop in the future.

Key Dates

22 March 1935	World's first regular television service opens in Berlin.
1945 - 1949	Allies supervise re-establishment of German broadcasting (initially radio only) on a public corporation basis in their zones of occupation.
5 August 1950	ARD founded to coordinate common interests of the regional corporations.
27 November 1950	First regular post-war German television service inaugurated on an experimental basis, broadcasting on alternate days, in Hamburg.
27 March 1953	ARD members agree to produce a common national television channel.
1956	BR becomes first corporation to introduce advertisements; others soon follow.
28 February 1961	'First Television Verdict' of the Federal Constitutional Court rules against Adenauer government's plan to introduce a commercial second channel on the grounds that a) broadcasting falls within the sole jurisdiction of the Länder, and b) a monopoly privately-run channel could not meet the Constitution's requirements for freedom of expression in the way that the public corporation channels do.
1 April 1963	The ZDF, established by the Länder in the wake of the 'First Television Verdict', begins transmission.
22 September 1964	BR opens the first regional 'Third Channel'; other corporations soon follow suit.
25 August 1967	Regular colour transmissions, using the PAL system, begin.
16 June 1981	Federal Constitutional Court concedes that technical developments (cable and satellite) now make possible the multiplicity of competing channels that is a precondition for freedom of expression if commercial broadcasting is to be introduced.
1 October 1982	CDU-led government, with a much more positive attitude to commercial broadcasting than its SPD-led predecessor, comes to power in Bonn.
2 January 1984	First German-language commercial television channel - 'RTL Plus' - begins broadcasting into Germany from Luxembourg; it then begins transmissions by satellite, where it is joined by its rival SAT 1.
4 November 1986	Federal Constitutional Court recognises that a 'dual order' (i.e. public and commercial) now exists in broadcasting, and declares that legislation must guarantee that the full range of public opinions has access to the airwaves.

24 March 1987	Federal Constitutional Court affirms that only the public corporations can realistically be expected to provide the full breadth and variety of programming, and therefore their continued existence must be guaranteed as a precondition for the admissability of commercial services.
3 April 1987	State Treaty between the Länder (in force from 1 December 1987) finally brings some order to the new organisation of broadcasting; the 'dual order' is affirmed, and the continued existence of the public corporations is guaranteed; satellite channels are distributed between the ARD, the ZDF, and the commercial operators, and between the Länder.
31 August 1991	State Treaty is updated (in force from 1 January 1992) to take account of Unification; the Eastern Länder have by now enacted their own Broadcasting Laws, thereby extending the existing West German system to the whole country.

Principal Organisations

ARD	Arbeitsgemeinschaft der Öffentlich-rechtlichen Rundfunkanstalten der Bundesrepublik Deutschland ('Association of German Public Service Broadcasting Corporations': the term is frequently used as the designation of the 'first' television channel)
B1	Regional television service of the SFB
BR	Bayerischer Rundfunk (Public-service broadcasting corporation [radio and television] for the Land of Bavaria)
3 SAT	Public-service satellite channel produced jointly by the ARD, the ZDF, and Swiss and Austrian television)
HR	Hessischer Rundfunk (Public-service broadcasting corporation [radio and television] for the Land of Hessen)
MDR	Mitteldeutscher Rundfunk (Public-service broadcasting corporation [radio and television] formed following Unification for the Länder of Saxony, Saxony-Anhalt, and Thuringia)
NDR	Norddeutscher Rundfunk (Public-service broadcasting corporation [radio and television] for the Länder of Hamburg, Lower Saxony, and Schleswig-Holstein, and - since Unification - Mecklenburg-Lower Pomerania)
N3	Regional television service produced jointly by NDR and Radio Bremen
ORB	Ostdeutscher Rundfunk Brandenburg (Public-service broadcasting corporation [radio and television] formed following Unification for the Land of Brandenburg)
RB	Radio Bremen (Public-service broadcasting corporation [radio and television] for the Land of Bremen)
RTL	Radiodiffusion-Television Luxembourgeoise (together with SAT 1 one of the two main commercial channels; now also broadcasts two further channels: RTL 2 and Super RTL)
SAT 1	Together with RTL, one of the two main commercial channels
SDR	Süddeutscher Rundfunk (Public-service broadcasting corporation [radio and television] for the northern part of the Land of Baden-Württemberg)

Channel Name	Date of Creation	Ownership and Funding	Viewers (1995)	Profile
Erstes Deutsches Fernsehen (ARD) Network of the eleven regional corporations: BR, HR, MDR, NDR, ORB, RB, SR, SFB, SDR, SWF, WDR	1953	Public corporation; funding derived from a mixture of licence fee (96%) and advertising (4%)	14.6%	General service; mixture of serious and popular programming covering virtually all genres: news, features, documentaries, shows, sport, films, etc.
Zweites Deutsches Fernsehen (ZDF)	1963	Public corporation; funding derived from a mixture of licence fee (80%) and advertising (20%)	14.7%	Similar to ARD (with which it shares some day-time programming), though generally regarded as slightly less serious
'Dritte Programme' (B1, BR, HR, MDR, ORB, SFB, WDR, N3, SW3)	mid-60s, except for the East German MDR and ORB founded after Unification	Regional programmes of the ARD corporations, six unique to one corporation (B1, BR, HR, MDR, ORB, WDR), two co-produced by adjacent corporations (N3, SW3); funding derived from licence fee (no advertisements)	9.7% nationally	Traditionally the most 'serious' channels, though all have now extended much more into 'mainstream' programming; regional output
RTL	1984	CLT Luxembourg: 47.9% Ufa (Bertlesmann 87%): 39.1% Westdeutsche Allgemeine Zeitung: 11% Deutsche Bank: 2% Funded by advertising	17.6%	General service, with heavy bias towards popular programming
SAT 1	1984	PKS (Leo Kirch): 43% Axel Springer Verlag: 20% Euromedia (Holtzbrinck Verlag): 15% Aktuell-Presse-Fernsehen (35% Axel Springer Verlag): 20% Otto Maier Verlag: 1% Neue Medien Ulm TV-Gesellschaft (Südwest Presse, Ulm): 1% Funded by advertising	14.7%	General service with heavy bias towards popular programming

Principal Channels

SFB Sender Freies Berlin ('Free Berlin Transmitter': public-service broadcasting corporation [radio and television] originally serving West Berlin, and since Unification the Land of Berlin as a whole)

SR Saarländischer Rundfunk (Public-service broadcasting corporation [radio and television] for the Land of the Saarland)

SW3 Regional television service produced jointly by SDR, SWF, and SR

SWF Südwestfunk (Public-service broadcasting corporation [radio and television] for the Land of Rhineland-Palatinate and the southern part of Baden-Württemberg)

WDR Westdeutscher Rundfunk (Public-service broadcasting corporation [radio and television] for the Land of North Rhine-Westphalia)

ZDF Zweites Deutsches Fernsehen ('Second German Television': public-service channel broadcasting a single programme nationwide)

Bibliography

ARD-Jahrbuch (annual)

Bausch, H. (ed.) (1980), *Rundfunk in Deutschland*, 5 vols, Munich, dtv.

Berg, Klaus (1995), 'Mischfinanzierung als Schutz vor politischer Einflu·nahme', *Media Perspektiven*, 3, 94-9.

Dähnhardt, Werner (1992), '"Mein Traum: ein Monopol"', *Der Spiegel*, 18 May.

Darschin, Wolfgang, & Frank, Bernward (1995). 'Tendenzen im Zuschauerverhalten', *Media Perspektiven*, 4,154-65 (156).

Ernst, Tilmann, & Hiegemann, Susanne (1992), *Privat-kommerzieller Rundfunk in Deutschland: Entwicklungen, Forderungen, Regelungen, Folgen*, Bonn, Bundeszentrale für politische Bildung.

Hiegemann, Susanne (1988), *Kabel- und Satellitenfernsehen: Die Entwicklung in der Bundesrepublik Deutschland unter ökonomischen, politischen und inhaltlichen Aspekten*, Bonn, Bundeszentrale für politische Bildung.

Humphreys, Peter J. (1994), *Media and Media Policy in Germany: The Press and Broadcasting since 1945*, 2nd edition, Oxford & Providence, Berg .

Media Perspektiven Basisdaten: Daten zur Mediensituation in Deutschland 1995 (1995), Frankfurt a.M. (supplement to Media Perspektiven).

Meyn, Hermann (1994) *Massenmedien in der Bundesrepublik Deutschland*, revised edition, Berlin, Volker Spiess.

Porter, Vincent, & Hasselbach, Susanne (1991), *Pluralism, Politics and the Marketplace: The Regulation of German Broadcasting*, London, Routledge.

Sandford, John (1976) *The Mass Media of the German-Speaking Countries*, London, Oswald Wolff.

Sandford, John (1985), '"Wollt ihr das totale Fernsehen?" The New Media Debate in West Germany', *Journal of Area Studies*, 12, 24-30.

Sandford, John (1995), 'The German Media', in Derek Lewis & John R.P. McKenzie (eds), *The New Germany: Social, Political and Cultural Challenges of Unification*, Exeter, University of Exeter Press, pp. 199-219.

Williams, Arthur (1976), *Broadcasting and Democracy in West Germany*, Bradford and London, Bradford UP and Crosby Lockwood Staples.

Television in Italy

Stephen Gundle

Prior to 1994 no European country had seen a broadcasting company turn itself into a political party and take power. Yet this is more or less what happened in Italy when the media entrepreneur Silvio Berlusconi founded his Forza Italia movement and forged the coalition with other right-wing parties that won the general election of March 1994. Although the coalition proved unstable as a governing alliance and Berlusconi was prime minister only for a short period (May-December 1994), his extraordinary rise provoked much alarmist comment. It seemed that a new phase of post-modern politics had been opened in which the power and techniques of the mass media could be manipulated by a small number of very rich and influential individuals to displace conventional political forces and take control of the institutions. According to this point of view, Italy was a laboratory of disturbing tendencies that could quite easily also come to fruition elsewhere.

In fact the whole Forza Italia phenomenon needs to be seen less as the harbinger of an all too plausible post-modern nightmare than as the outcome of a set of highly specific relations between television, politics and society (see McCarthy, 1996 and Gundle and O'Sullivan, 1996). In Italy, the development of visual broadcasting coincided with a far-reaching set of social changes which, together, saw the country pass from a basically agricultural economy in the 1950s to an industrial one in the 1960s and a post-industrial one in the 1980s. Because there was little shared culture or sense of belonging prior to television, this immediately assumed a central place in the organisation of society. Television has been credited in Italy with having assisted linguistic unification, overcome particularism, spread consumerism, facilitated the emergence of national retail distribution networks, raised the level of political awareness, and brought even the most marginal social groups and geographical areas into the national community. However, the dynamic, innovatory role of television was tempered, on the one hand, by the way it always sought to incorporate traditional, conservative values (family, religion, localism etc.) and, on the other, by close political supervision and control. From the Fascist period through to the present, there has always been a high level of interaction between the broadcast and print media and the political sphere. For most of its history, television, like radio before it, has been conceived as a tool of an established political elite. When this elite collapsed amid numerous corruption scandals in 1992-93, few were surprised that the country's leading commercial broadcaster emerged to offer himself as a national saviour.

The aim of this chapter is to illustrate the specific characteristics of Italian television and to examine the nature of its social and political impact. The history of television will be outlined first, then legislation, the role of government, regulation and advertising will be discussed. The third and fourth parts of the chapter are dedicated to an analysis of the various public and private networks and the broader relationship of television to society.

The history of Italian television

The history of television in Italy can be divided into four phases. The first phase

(1954-74) was marked by the monopoly of the state broadcasting company RAI (initially Radio Audizioni Italia, from April 1954 Radiotelevisione Italiana). The second phase (1974-84) saw the development alongside RAI of an extremely varied and virtually unregulated private sector. The third phase (1984-1993) saw the consolidation of a broadcasting 'duopoly' consisting of RAI and Berlusconi's Fininvest (now Mediaset) group, which emerged as the dominant player in the independent sector following a process of concentration. The current phase (1993–) is marked by a struggle (thus far not concluded) between those who wish to preserve the existing balance of power and reformers who want to change the system to allow greater pluralism and diversity.

Let us examine each of these phases in turn. Following experimental broadcasts in Turin in 1949 and at the Milan trade fair in 1952, regular transmissions began in the North of Italy in September 1953. The date normally given for the start of television broadcasting, however, is 1 January 1954, the day of the official inauguration. Programme output showed a desire to imitate the BBC's aim 'to inform, educate and entertain'. According to Franco Monteleone, the leading historian of Italian radio and television, the structure of the schedules bore witness to the pedagogical approach of early broadcasters who were determined to avoid the commercial, entertainment-dominated 'flow' of American television (1992: 279; for an alternative view, see Wolf 1989: 52-3). Instead they wanted to use the medium to mould the beliefs and values of the population. Schedules were organised around regular weekly appointments, with Monday evenings being dedicated to films, Fridays to theatre, Saturdays to variety etc.). This, however, did not mean that RAI was above making use of American formulas in devising popular programmes. These were invariably heavily adapted, however, to take account of different tastes and expectations. The best-known example of this is *Lascia o raddoppia?* (Double or Quit?), a quiz programme with money prizes that was a local version of the hit US show *The $64,000 Question* (see Buttafava, 1980 and Gundle, 1986).

Aided by the government's decision to promote the rapid expansion of broadcasting and the enormous success of *Lascia o raddoppia?*, television fairly quickly became a fixed part of Italians' lives. It appealed so widely and massively for three main reasons. First, it was an extraordinary novelty, a window on the world and a source of information about modern, urban styles of living that was highly valuable at a time of vast social and economic change. Second, it cost little. Although relatively few families could afford to buy a set in the 1950s, free viewing in public places such as recreational centres and bars (where the only cost was the price of a drink) was very widespread. In this way, television reached sectors of the population, such as the old and infirm, the poor and those living in isolated mountain and rural areas, who had never even been to the cinema. Third, it cloaked modernity in the familar guise of tradition. Catholic values were on the whole respected, family entertainment was emphasised and the national passions of sport and popular song were given due space. Presenters and link announcers were the sort of cordial, courteous, modestly attractive people that anyone would be pleased to receive in their home (Gundle, 1996b).

For nearly two decades the RAI formula of public service television worked well. Although there were complaints from opposition parties about political discrimination (news broadcasts were unfailingly pro-government and current affairs coverage was minimal before the 1960s), the carefully balanced output was generally appreciated. The inauguration of a second channel in 1961 marked the coming of age of visual broadcasting and heralded the beginning of a season in which RAI enjoyed a reputation as one of the world's foremost broadcasting organisations. Its standing was particularly high in the area of sports coverage following the 1960 Rome Olympics.

The second phase was a turbulent one in which RAI lost its monopoly and was forced to come to terms with an emerging private sector. Although the political elite agreed on a substantive reform of RAI in 1975, the Constitutional Court issued a judgement the following year which legalised private terrestrial broadcasts at local level. Following the initial blossoming of hundreds of local stations, some of which were commercial, others voluntary, major players moved in and began to create the premises of national commercial networks. The publishers Mondadori, Rizzoli and Rusconi sponsored stations, as did entrepreneurs including the Agnelli family, Silvio Berlusconi and the dairy magnate Camillo Tanzi. The new stations were very different from RAI because they relied heavily on films and imported TV drama series, soap operas and cartoons. Although they could not broadcast live, they appealed because they were upbeat, fast-moving and on the whole not paternalistic.

The third phase saw the consolidation of Berlusconi's control over the private sector. Using American broadcasting techniques and a very sharp approach to advertising sales, he made his Canale 5 network the leader in the field and forced most of his rivals to quit. By buying the failing channels Italia 1 and Rete 4, respectively from Rusconi and Mondadori in 1983 and 1984, he built an empire that frequently took over 50 per cent of national audiences. Local stations and small networks were rapidly thrust to the margins of the broadcasting system. Berlusconi challenged RAI by creating a dynamic on-screen atmosphere, poaching established stars and, last but by no means least, by constructing a network of supporters among the political elite, most especially the small but pivotal Socialist Party (PSI), whose leader, Bettino Craxi, was prime minister between 1983 and 1987 (Martini, 1990; Gundle, 1996). The cut-throat competition between RAI and Berlusconi's Fininvest empire coincided with, and in part was responsible for, important changes in the nature of Italian television. In the 1980s it took on daily rather than weekly rhythms, it became more geared to the logic of industry and competition, and ratings became a primary concern (Wolf, 1990).

In this phase of Italian television, RAI struggled to maintain its preeminence and the loyalty of audiences. Political influence over it grew and became more diversified following the institution of parliamentary control in 1975 and the creation of a third public service network in 1979. The company hit back at the competition by buying in more foreign material, engaging in European co-productions, making use of its monopoly of national live broadcasting, and investing in prestige productions of biographies of national heroes (Marco Polo, Verdi) featuring partly foreign casts (Richeri, 1990).

In 1990, after much delay, a law was passed which introduced elements of regulation to the private sector. This legitimised Berlusconi's ownership of three national networks. As a result, in 1991 Canale 5, Italia 1 and Rete 4 were able to begin broadcasting live and develop proper news services. However, after 1992 a fourth phase appeared to have been opened. The 1990 law was widely questioned as the old political elite fell from grace, the Constitutional Court urged action to permit more pluralism in the private sector, the innovations of satellite and cable were threatening and there was talk of privatising part or even all of RAI. Pressures for reform increased when Berlusconi entered politics and was subsequently incriminated, along with many of his senior employees, on charges of corruption and tax evasion. In the summer of 1996 it was still too early to say what the future would hold, but there were some signs that the so-called 'duopoly' of RAI and Fininvest might soon be a thing of the past.

The politics of television

The legislative framework

Television was born and grew up within the democratic context of the postwar Italian Republic. The legal and institutional framework of television, however, was that which pre-existed for radio, which had taken shape during the Fascist era. At the end of the war, the state-dominated broadcasting company EIAR (Ente Italiano per le Audizioni Radiofoniche) changed its name to RAI, but there were few changes of personnel or structure. 'After the decentralisation and the short-lived experience of democratic management of 1945', David Forgacs has written, 'a system emerged which bore remarkable similarities to that which had obtained under Fascism, minus the Fascist information content' (1990: 108–9).

Under this system, RAI was not directly owned by the state. The board included representatives of private industrial interests, the state telephone company SIP and the state industrial holding company IRI, as well as government nominees. Although the government quickly established its predominance, industrial factors were always important in RAI's development. RAI's monopoly of broadcasting was confirmed in a presidential decree of 1952 which approved a convention between the company and the Ministry of Posts for twenty years. However, both the specified temporal limitation of the convention and article 21 of the republican constitution, which guaranteed freedom of expression 'by word, writing or any other means of communication', set the seed of doubt about its legitimacy.

The Constitutional Court played a crucial role in undermining the public monopoly. In 1960 it confirmed the constitutional legitimacy of the monopoly on practical grounds only and, even then, on condition that citizens were given the opportunity to express their opinions within the monopoly and that news coverage was objective. In the late 1960s and early 1970s both democratic movements and private commercial lobbies called for the abolition of the monopoly, as the exclusive concession came up for renewal in 1972. They based their arguments variously on democratic and commercial grounds, as well as practical ones of technological advance since the old argument about the limited number of available frequencies no longer applied. There were two responses to this. The first was the law on reform of RAI of 1975. This confirmed that broadcasting was a public service and that it would remain a state monopoly, with the exception of local radio and cable television, which had been legalised by a Constitutional Court ruling of 1974. The law transferred formal control of RAI from the government to a parliamentary commission which nominated ten of sixteen members of a board of governors. The Court supplied the second response when it modified the main premise of the 1975 law. The judges declared that RAI's monopoly was justified at national level only. The way was thus opened for private broadcasters to transmit locally (for more detail, see Forgacs, 1990: 141–45).

Between 1976 and 1990 the private sector remained theoretically unrestricted, although in fact, as soon as he had won a monopoly position, Berlusconi joined forces with political allies to prevent the entry into the market of any potential substantial private competitors. In response to local magistrates, who had started ordering the blackout of Canale 5 in their regions on the grounds that its transmissions were not purely local, the Craxi government issued in 1984 a decree (subsequently converted into law in February 1985) which provisionally granted to existing stations only the right to broadcast beyond the local ambit until such time as a general regulatory law was passed. The Constitutional Court repeatedly called for anti-trust legislation and for regulation of the sector, but this advice went unheeded by coalition governments which included Berlusconi's allies. Private broadcasters lobbied extensively and successfully against any restrictive legislation.

The law of 1990 was backed by the very parties that had previously blocked legislation. They changed their position for the sole reason that the Constitutional Court threatened to intervene autonomously if parliament persisted in not doing so. Yet the law did not really restrict private trusts as it fixed a maximum limit of three networks per operator and introduced only minor forms of regulation. For this reason it was described as being tailor-made for Berlusconi. Only in 1992 did the government establish, under the terms of the law, which networks would be given official recognition as national stations (three RAI, three Fininvest, Videomusic, TeleMonteCarlo, Rete A).

The crisis of the political system from 1992 witnessed moves to reform the 1990 law, which was seen as a product of the old system. In 1993 a 'mini-reform' of RAI was passed that depoliticised the management of RAI and urged financial restructuring (Mazzoleni, 1994; Gundle and O'Sullivan, 1995). This was intended as a first step towards a more general reform of the whole broadcasting system. In fact moves towards general reform were delayed by Berlusconi's victory in 1994. The issue returned to the top of the agenda following the victory of the centre-left Olive Tree coalition in the 1996 election. In keeping with the Constitutional Court's insistence that no private individual should be allowed to operate more than two channels, a bill was set before parliament in July 1996. In the event of this becoming law, both RAI and Mediaset will be obliged by August 1997 to divest themselves of one network: one RAI network will become federal, while one Mediaset network will be transferred to satellite or cable. In addition, no single subject will be allowed to control more than 20 per cent of terrestrial television programmes.

Links with Government

Television in Italy has always been politically dominated. The ruling Christian Democrats (DC) ousted left-wingers from RAI in the late 1940s and did their best to stamp early television with their imprint. The appointment as the first director-general of Filiberto Guala, a state sector manager linked to the ruling party, signalled the desire of the politicians to take control (Chiarenza, 1978: 69). A fervent Catholic, he saw television as a 'new pulpit' to be placed at the service of the political majority. However, control in the early years was not easily achieved as many long-standing RAI employees resisted attempts to undermine their autonomy. The DC's allies in the political centre also constituted an obstacle. A long struggle was concluded towards the late 1950s, when the DC finally adopted a conciliatory attitude towards lay and even moderate left-wing forces. Within RAI, the shift from integralism to compromise was sealed with the appointment as director-general in 1961 of Ettore Bernabei, a DC loyalist of a more modern and subtle cast than Guala and his successors. Under his leadership, which lasted until 1974, RAI remained an institution in which DC influence was very marked; but there was also space for a range of other views and a company ethos was allowed to prosper.

In the early days there was very little politics on air, apart from pro-government news programmes. It was feared that any other form of coverage would play into the hands of the opposition, which was led by the Italian Communist Party. This situation changed in the 1960s, when the government coalition was widened to include the Socialists. Party political broadcasts began in 1960, current affairs increased, and documentaries and films relating to recent Italian history (especially Fascism and the Resistance) were shown. At the same time, the Socialists gained a foothold in RAI.

The big change occurred in the aftermath of the 1975 reform, which marked the transition from executive domination to parliamentary control. Formally, it was no longer the government that ruled RAI but the parties represented in parliament. As a result, each party was able to insert key people in decision-making positions with influence over programme policy, scheduling and so on. The DC and PSI took the lion's share, but other parties, including the Communists, also received posts. The spoils system (*lottizzazione*) was rendered explicit in 1987 when each of the three RAI channels became the fief of one of the major parties, the DC taking the flagship RAI-1, the PSI RAI-2 and the Communists the Cinderella network RAI-3. The main influence was at the level of appointments but output was also affected. News broadcasts were quite openly skewed towards the controlling party and bias was apparent in current affairs too. Politicians also vied to appear on chat shows and variety programmes.

Peppino Ortoleva has argued that between 1974 and 1994 there was a complete fusion between the political system and the broadcasting system. Political colonisation was so complete that television simply became an adjunct of the party system (Ortoleva, 1995: 10). In this scheme Fininvest's designated function was to provide low level depoliticised entertainment to keep the masses happy. There is much evidence in favour of this argument: the massive political interest in the media, the growing importance of television in political communication and mobilisation (see Pasquino, 1983 and Gundle, 1992), the proliferation of current affairs and the 'party bulletin' nature of news broadcasts. But there are also objections. It underestimates the extent to which politicians, in a phase of decline in party organisation, were in thrall to the media. They did not really understand television and were fearful of its influence. The Communists, for example, refused to accept commercial television before the mid 1980s (Gundle, 1995: 473-79). Moreover, political control came up against the fact that industrial concerns and technology were also driving factors in television and so, increasingly, were economic ones.

The 1990s saw growing concern about excessive politicisation and the effects of *lottizzazione*. The main result of this was the 1993 mini-reform of RAI. Under this reform, the presidents of the two chambers of parliament were to nominate five people of 'renowned independence' to run RAI. At first a technocratic board was appointed (see Gundle and O'Sullivan, 1995) but, following the victory of the right in 1994, this was replaced by a new board consisting of formally independent persons who in fact were aligned with the right (Gundle, 1996). These resigned or were dismissed when the centre-left took power in 1996. The latter behaved a little better than the right insofar as its appointments were both more prestigious and more balanced. But the new team remained subject to political influence, proving that the new law was insufficient to break established habits of political control. The centre-left's reluctance to promote the complete depoliticisation of RAI is, however, understandable so long as the leader of the opposition controls such vast resources and regards his media empire as a political tool.

Regulatory mechanisms

There are two ways of looking at the question of regulation. From one point of view Italian television is very closely supervised. The longstanding tradition of political involvement has meant that broadcasters have never enjoyed the degree of professional autonomy that is taken for granted in some countries. As a result, television has rarely represented public opinion accurately or acted as a stimulus to improvement in a way that is usually held to be a normal function of the mass media in democratic societies. The director-general of RAI or, following the 1993 mini-reform, the organisation's president, has always been expected to intervene to censor managers, editors or journalists who offend powerful interests. When internal disciplinary procedures prove insufficient, the parliamentary watchdog commission can be relied upon to demand direct political action.

From the other point of view, and in part because of the lack of interest of the political elite in formal regulatory mechanisms that might limit its own influence, Italian television is very underregulated. As has been shown above, 'for ten years a perverse web of politics, business and finance blocked any hypothesis of a correct and harmonious solution to the problems born of the rapid explosion of the media system in Italy' (Monteleone, 1992: 472). Using their influence, powerful interest groups were able to secure delays, compromises and partisan decisions. The Italian state in this sense cannot be said to have ever really acted as an effective umpire or regulator. When they came, rules were fixed late and tended simply to ratify states of affairs that had come about entirely spontaneously (Monteleone, 1992: 529–31).

Proper, if still relatively weak, regulation was introduced under the 1990 law, which created the figure of the Guarantor for Broadcasting and Publishing. The establishment of an independent authority to integrate the sort of control exercised in the past solely by parliament and the government matched trends elsewhere. But because there was no scaling down of, for example, the power of the Ministry of Posts in awarding concessions and applying sanctions, or of the Treasury Ministry in controlling IRI's stake in RAI, there was a confusion of roles. The Guarantor was not able either to act independently or to govern the sector in any meaningful way (Caretti, 1995: 303–306). Nevertheless, the Guarantor was seen as a figure of some importance and the office has received new powers from time to time. Under a 1993 law governing electoral propaganda, for example, the Guarantor became a sort of campaign referee whose job was to ensure equality of access for all parties on public and private channels (Bettinelli, 1995:124–40). The situation is likely to become more rather than less confused if the 1996 bill becomes law as, without redefining the functions of the Guarantor or reducing those of government departments, it envisages the creation of a new regulatory body whose members will be elected by parliament.

The powers of the Guarantor do not extend in general to questions concerning programme content. Sexually explicit material, for example, is regulated by a section of the penal code dealing with offences against public decency. Just what decency means, however, has been the subject of much redefinition. In the 1950s Catholics succeeded in securing the covering-up of the legs of dancers but, by the late 1970s, a total revolution had occurred in public standards of decency and private stations were able to broadcast pornographic films without fear of penalty. The image of Italian television as consisting largely of housewives stripping and nude game shows is of course false, but on some local channels, especially in Rome and the South, virtually anything goes. Mainstream channels are more circumspect, but in the late 1980s 18-rated films such as *Last Tango in Paris*, banned even in cinemas for many years in Italy, were shown in mid-evening. Moreover, a number of shows have featured stars of Italy's blue movie industry such as Cicciolina and the late Moana Pozzi, while one or two others have been fronted by transsexuals. Although displeasure is sometimes voiced at such developments, on the whole an easy-going, *laissez-faire* atmosphere prevails.

Advertising

One area which has always been subject to strict regulation is advertising on public service television. In the early years (advertising began in 1957) it was limited to a maximum of five per cent of the overall broadcasting ouput. This was as a result of pressure from publishers and film producers and exhibitors, who were fearful that their own revenues would be damaged by the new medium. The management of advertising was placed in the exclusive hands of a joint IRI-RAI society, SIPRA, which had been created to perform a similar function for radio before the war. Because the demand for advertising space outstripped supply by approximately six-fold (Chiarenza, 1978: 91), SIPRA performed an important gatekeeping function and acted as an instrument for the distribution of favours to allies of the DC in industry.

The nature of advertisements was also closely regulated because political and cultural elites were hostile to naked commercialism. Programmes were not interrupted by spot announcements; instead advertisements were grouped together into a programme of their own known as *Carosello*, which was broadcast every evening after the news. Within these, there were strict limits on how often a product could be mentioned, with the result that advertisers were compelled to invent stories and characters (such as the cardboard Mexicans Caballero and Carmencita for Lavazza coffee) or develop plots using stars well-known already to the audience. Intended to eliminate the vulgarly commercial from advertising, the *Caroselli* in fact became immensely popular. The *Carosello* system came to an end in 1977. But close regulation continued. RAI's advertising ceiling was set by the government, while, under the 1975 law, it fell to the Parliamentary Watchdog Commission to determine annually the maximum limit of permitted advertising revenue. These constraints were a godsend to commercial broadcasters for, unlike RAI which also received revenue from the licence fee, they relied entirely on advertising income for funding. Subject to no constraints at all, private stations could harness as much revenue as they were able to muster and broadcast as many advertisements as the audience would tolerate (Forgacs, 1990: 182; Bruzzone, 1981: 598-99). At local level, there are still a large number of stations which are either entirely devoted to tele-shopping and sales or which feature it heavily. Quack remedies and potions, furniture, jewellery, carpets and fitness equipment are all offered for sale by loud, aggressive types and some colourful personalities. The goods often give bad value for money; their appeal derives exclusively from their having appeared on television.

In the 1980s, the private television networks broadcast vast numbers of spot advertisements, interrupting films and programmes every five minutes and broadcasting advertising louder than usual transmissions. By 1984 the quantity of advertising on private television was ten times that on RAI-TV (Forgacs, 1990: 184-85). The massive increase was largely due to Berlusconi. His success in the private sector derived less from the quality or appeal of his programmes than from his ability at selling advertising space often by ruthless discounting and undercutting. He also succeeded in bringing in a range of middle-sized companies that previously would not have considered television advertising (Forgacs, 1990: 179-87). By the mid-1980s, the advertising arm of the Fininvest group, Publitalia, had become the most powerful force in the advertising industry.

The triumph of a commercial logic in the private sector revolutionised the system. Programmes, schedules, tone and even the appearance of on-screen personnel were all determined by the need to satisy the needs of advertisers. Saturation led to avoidance tactics by audiences now skilled in the use of the remote control. As a result some self-regulation was adopted to limit the overall volume of advertising and move to forms of sponsorship and in-programme promotion. On both commercial channels and RAI, which copied the tactic with relatively little hesitation, presenters break off to do sales turns that have become shows within a show. Sports events are interrupted by near-subliminal flash advertisements.

Ortoleva has drawn attention to the importance of colour to commercial television. Colour broadcasting started in Italy in 1977 and was used most effectively by private television to create a parallel universe of consumerist desire in which everyone was beautiful, happy, healthy and prosperous. This was 'dream television, that owes its success with the audience to its detachment from any meaningful reference to the concrete reality of life' (Ortoleva, 1995: 15). Self-referential and self-enclosed, it constitutes a world-unto-itself that is surreal, 'but which can be transformed in part into actual life through acts of consumption' (Ortoleva, 1995: 16).

The importance of commercial considerations was formally acknowledged in 1986 with the creation of Auditel. Founded jointly by RAI, Fininvest and advertisers, this organisation measures ratings by means of a device that is inserted in the TV sets of sample families. The results then determine advertising charges. Despite this measure of institutionalisation, the vast increase in advertising led to calls for a ceiling and for limits on the frequency with which films especially could be interrupted. Because private interests and their allies in parliament objected, nothing was done on this score until the 1990 law fixed ceilings for RAI (12 per cent of output) and Fininvest (18 per cent) as well as more specific constraints. A referendum which aimed to ban outright advertising breaks in films, opera and so on was defeated in 1995 following a strenuous campaign against it by Fininvest. Under the terms of the 1996 bill, the Mediaset ceiling would be lowered to 16 per cent or, should the group prefer it, to 12 per cent on one network only.

The broadcasting networks

As we have already seen, television is dominated by public television (RAI-1, RAI-2, RAI-3) and one large private group, Berlusconi's Mediaset (Canale 5, Italia 1, Rete 4). In addition, TeleMonteCarlo is a significant minor player. It is these channels which will be described in this section, not the smaller syndicated channels or local stations, which nonetheless continue to play a role. It is appropriate to begin with Mediaset, because the impact of Berlusconi has been so widespread in television over the last fifteen or so years. Although RAI conserves its public service role, it has fundamentally changed many of its practices in the context of competition.

Canale 5

Canale 5 was Berlusconi's first nationally syndicated channel and it remains his flagship. By the early 1980s it had won an audience share of around 50 per cent due to Berlusconi's use of American-style daily scheduling (rather than the traditional RAI schedules which privileged weekly appointments). *Dallas* and *Dynasty* were early strong suits in the war with competitors and heavy use of imported material still continues. Films feature prominently in schedules, as do quiz shows and variety programmes feature prominently too. Berlusconi lured veteran presenter Mike Bongiorno from RAI in 1980 and since then he has achieved high ratings with regular quiz and game shows. Other stars, including Pippo Baudo and Raffaella Carrà, were poached from RAI in 1987 but their migration was not a success and they soon returned to state television. A typical Canale 5 success of recent years is *OK il prezzo è giusto*, a local version of *The Price is Right* which is fronted by the Italian Cilla Black, Iva Zanicchi. For many years, the late evening slot has been dominated by the daily talk shows of the portly journalist Maurizio Costanzo. His guests are occasionally well-known, but particularly favoured are eccentrics and oddballs. Since Fininvest began live broadcasts in 1991, Canale 5 news has won a reputation for American slickness and brevity as well as impartiality. The look of the channel is glossy and opulent on the whole, but the tawdry and the tacky are also present.

Italia 1

Italia 1 is the youth-oriented network of the Mediaset stable. Although it also transmits mainstream films and variety shows, it is best known for two programmes, the popular Sunday evening comedy show of the 1980s *Drive in* and the notorious weekday afternoon programme featuring dancing teenage girls *Non è la Rai* (It's not the RAI). Invented by Antonio Ricci, the former consisted of two hours of rapid-fire gags provided by a team of comics who played 'types' instantly recognisable to a youth audience (the over-age southern student of Milan's Bocconi University, the paninaro enamoured of all things fashionable and foreign etc.). Supposedly set in a drive-in cinema (a phenomenon unknown in Italy), the regular cast was complemented by a team of scantily-dressed 'fast-food girls' and a standing audience paid to supply laughter (usually of the forced variety). The real innovation of the show was its construction of a model of entertainment whose rhythms were exactly those of the advertisements that, with great regularity, interrupted it. In the 1990s Italia 1 has featured, in addition to the pedophiles' delight *Non è la Rai*, the daily singalong show *Karaoke* and *Striscia la notizia*, a brief satirical broadcast parodying the news.

Rete 4

Originally owned by Mondadori, Rete 4 is today a channel which broadcasts a heavy diet of telenovelas from Brazil and Mexico and North American soaps. These are great favourites with housewives and students. It would be quite normal for an entire day's programming to consist of imported television films, save perhaps an old Italian film broadcast in the evening. Its director of news, Emilio Fede, regarded as more *berlusconiano* than Berlusconi himself, has given the channel's news broadcasts an unenviable reputation for bias and partisanship. The weakest of the channels in the Berlusconi stable, Rete 4 regularly attracts an audience share of only 7 per cent. In the event of its owner being compelled to divest himself of a channel, Rete 4 would almost certainly be the one he would choose to sell or move to satellite or cable.

RAI-1

The best-funded and most prestigious of the three RAI channels, RAI-1 has always had an air of officialdom about it. Its news and current affairs programmes typify the glum sanctimoniousness that many find so unappealing about public service television. Dominated by the Christian Democrats and, since their demise, by a less clearly-defined Catholic interest group, the channel competes directly with Canale 5, mainly on the terms and terrain of the latter. In the past, it offered a variety of programmes of different genres, including adaptations of literary classics, but in recent years it has largely abandoned strategic planning in favour of cheap and cheerful variety shows (cheap, that is, in tone and appearance, not actual cost which can be exorbitant). A great favourite is the Sunday afternoon container show *Domenica in*, which provides several hours of filler material (games, phone-ins, horoscopes, book, film and record presentations, songs, magic etc.) which is of little interest to anyone save perhaps the old and the sick. In the 1980s RAI made several foreign language mini-series of the lives of Italian heroes (Marco Polo, Columbus, Verdi) but it did not succeed, as it had hoped, in selling these widely abroad. The long-running Mafia saga *La Piovra* (The Octopus), of which there have to date been seven separate mini-series, is currently the jewel in its crown, although several episodes have provoked intense controversy (Gundle and O'Sullivan, 1996: 207-8). RAI-1 has increasingly been seen as beached whale tied to an outdated type of paternalistic 'generalist' broadcasting, but it still regularly attracts large audiences. It comes into its own when it broadcasts events that arouse great public interest like the annual San Remo song festival and international football matches.

RAI-2

For quite some time, RAI-2 has been seen as a potential BBC-2 to RAI-1's BBC-1. Television reformers and some insiders have argued that it should be slightly more high-brow than RAI-1 and offer less immediately accessible programmes. In fact this has scarcely occurred. Only the channel's long-running late evening current affairs transmission *Mixer* could be considered challenging and informative. Even this was hamstrung for many years by the need to kowtow to the Socialist Party and its leader Bettino Craxi. In keeping with the oft-trumpeted desire of its political masters to become the dominant political force in the country (despite a share of the vote that never rose above 14 per cent), RAI-2 aimed to compete across the board with RAI-1, duplicating its news services, variety shows and so on. A huge success has been the North American soap *Beautiful* (The Bold and the Beautiful), daily broadcasts of which mesmerised the country for five or so years. Since the demise of the Socialists (see Gundle, 1996), the channel has struggled to redefine its identity. The appointment as director in 1996 of Carlo Freccero, a scheduling genius who masterminded the early successes of Canale 5 before falling out with Berlusconi and migrating to Paris, may herald a turning point.

RAI-3

For nearly ten years, RAI-3 limped along achieving only minimal audiences. Created mainly for political reasons as a sop to the the Communist Party, it was intended that it should reflect the regional diversity of the country. In fact it was so underfunded that it could not even be picked up in all areas. A significant change in the fortunes of RAI-3 took place in 1987 with the appointment as director of the left-wing literary critic Angelo Guglielmi. Virtually single-handedly, Guglielmi invented a type of *television-verité* that was completely new to Italy. Lacking funds to make drama or other costly programmes, he pioneered shows which spectacularised real situations and problems such as court cases (*Un giorno in pretura*) and missing persons (*Chi l'ha visto?*). Other landmark shows included the highly tendentious current affairs transmission *Samarcanda* and the deliciously surreal compilation show *Blob*. The latter offered each day a selection of the worst horrors (gaffes, vulgarity, technical faults, heated exchanges, exhibitions of ignorance) broadcast on all channels in the previous 24 hours. A by-product of the remote control and channel-hopping, *Blob* dwelt gleefully on the very worst of Italian television and, in so doing, offered of it a uniquely persuasive immanent critique that was entirely in keeping with the character of contemporary broadcasting. To complete the channel's unconventional profile were its news broadcasts which, while they were certainly more lively and alert to civil society than those of RAI-1 and RAI-2, were no less politically biased - in this case towards the Communists (and the PCI's successor, the Democratic Party of the Left) rather than the Christian Democrats or the Socialists. Between 1993 and 1996, RAI-3 faced a consistent attempt to 'normalise' it (see Gundle, 1996a: 213-14), which began with the dismissal of the director of news and continued with the removal, in 1994, of Guglielmi. Although RAI-3 was seen to have played a vital role in mobilising public opinion against the old political elite in the early 1990s (Gundle and O'Sullivan, 1996), the refusal of seasoned staffers to accept a more conventional view of broadcasting in the aftermath of the crisis ensured that they had few, if any, political allies. Under the proposed reform announced in 1996, RAI-3 would become a federal network, although, at the time of writing, it is unclear precisely what this would mean.

TeleMonteCarlo-Videomusic

TeleMonteCarlo (which, despite its name, now broadcasts from Rome) was one of the earliest commercial stations. By virtue of its headquarters being abroad, it enjoyed advantages denied its competitors (including live news broadcasts). Although audiences have always been small (today the average share is just three per cent), one or two programmes have struck wider chords. The football talk show *Gala gol* launched the showgirl Alba Parietti, who is now one of Italian television's biggest personalities. The early evening magazine programme *Tappeto volante* (Flying Carpet) also has a certain following. Ownership of the station has changed considerably as different interests, ranging from Brazil's Rete Globo to the Fiat group, have taken part shares and considered turning it into a third major broadcaster. The current proprietor is Vittorio Cecchi Gori, a sort of Berlusconi in Tuscan sauce who is the country's largest film producer as well as owner of the Fiorentina football team and a senator for the Olive Tree coalition. In a recent hostile takeover, Cecchi Gori also gained possession of Videomusic, a niche popular music broadcaster with a mainly youth audience. The music video output of this station has been reduced as part of a plan to turn the station into TeleMonteCarlo 2.

Programme Production

Italian television lacks many programmes that are considered staple parts of the schedules in Britain, America and other countries. There are no long-running soap operas (although there have been some short-lived attempts to create them, including Italia 1's youth-oriented *I ragazzi del muretto*). There are no situation comedies (save *Casa Vianello*). Very few documentaries are broadcast. Moreover, there is very little drama, a genre which was quite popular in RAI in the 1960s. Instead there is an extraordinary quantity of imported material, mainly from the United States but also from South America and Japan. In the 1980s, for reasons connected with deregulation, Italy became the principal world importer of both American and Japanese television. In addition, there are innumerable quizzes, game shows, people shows and lengthy discussion programmes. The genre which dominates television entertainment, however, is the variety show.

Typically, a variety show will be fronted by a star presenter or well-known comedian who will be accompanied by one or two attractive soubrettes (frequently one blonde and one brunette) who will wear a selection of revealingly ostentatious gowns and smile profusely. The more talented will exhibit themselves in a series of dance sequences accompanied by a *corps de ballet*. The show will additionally consist of a series of more or less funny sketches, some songs, a phone-in game, and some guest slots as well as the obligatory product promotions. Such programmes were typical Saturday evening fare in virtually all countries in the 1950s and 1960s, but they often did not survive the 1970s. In Italy, by contrast, variety has prospered and multiplied.

Several explanations may be offered. First, for complicated reasons which cannot be discussed here, the sense of seriality and continuity which is a necessary premise of soap opera and sitcom production is not very developed in Italian culture. Second, regular management changes and political interference have periodically impeded strategic planning and investment. This situation was exacerbated by the enormous debts accumulated by both RAI and Fininvest in the 1980s (see Gundle, 1996: 197-200). Third, variety-style theatrical entertainment was extremely popular in Italy and, although it declined from the late 1940s, it conserved a following well into the 1970s. Consequently it influenced other forms of cultural production to a greater extent than was the case in some countries. Finally, variety shows can often be laid on with little notice (if not in-house, then by contracting out) and can provide an assortment of elements which can fill out container and game shows, discussion programmes and so on. Inoffensive and appealing to the least sophisticated palates, variety offers the sort of light-weight escapism which Italy's political and economic elites have traditionally preferred to supply to the population.

Television and society

Since the late 1970s, television has been the dominant information and entertainment medium in Italy. Today, much of the periodical press, the music industry, cinema and sport enjoy a close relationship with public and private television. Much of political life also takes place in television studios. This is not to say that the link between television and these spheres is always a harmonious one. On the contrary, the inequality of the relationship often produces friction. This is especially marked in cinema and the press. Through the 1960s and 1970s Italian cinema enjoyed a very high world reputation as well as comparatively buoyant domestic audiences. Its quite sudden decline, although by no means solely due to the explosion of commercial television, was nonetheless blamed on this (Gundle, 1990: 220-22). Directors campaigned against advertising breaks which were allegedly destroying the artistic integrity of their works and in a few cases made films overtly hostile to television (see Fellini's *Ginger and Fred* or Nichetti's *The Icicle Thief*, for example). However, by the late 1980s such complaints dwindled as the various public and private networks increasingly became involved in film production.

The attitude of the press is similarly contradictory. A vast amount of space in the daily and weekly press is given over to questions concerning television. These range from political disputes about appointments or management policy to the behaviour of moderators of current affairs broadcasts and the pay of entertainers. The strength of the star system of the small screen is demonstrated by the extraordinary interest that is displayed in the family and love lives of personalities, their tastes and their homes, and their opinions about virtually every issue (see Gundle, 1996b). Yet, at the same time, the daily press adopts a superior attitude towards all things televisual and even the weekly gossip magazines (*Novella 2000, Eva tremila* etc.) ridicule television personalities. Dependency, it would seem, produces not only familiarity but also contempt.

One of the features of television which most frequently gives rise to complaint is its self-referentiality. At one time, television looked out at an external reality. It brought the world to the local bar or to the living room. It never did this unselectively or neutrally, but basically this was its main function. Since the late 1970s, television has, by contrast, been overwhelmingly concerned with its own history, concerns, events and stars. This state of affairs, which Umberto Eco has described as 'neo-television' (1990), is a product of the growing role of television in people's lives, its expansion as an industry and the fact that today there is no 'reality' which has not either gazed at the mass media or been the subject of the latter's gaze. Neo-television is not, of course, an exclusively Italian phenomenon. But in Italy it does take on special connotations because of the unusual importance of the medium in creating a shared sense of belonging and developing a collective identity. It also has the dubious advantage of reinforcing the historic vocation of television to cover over and ignore forces and problems which the dominant elite finds irksome.

For a majority of Italians, television has served above all as 'a mirror... in which they have seen their aspirations realised and their convictions respected.... Rather than an agent of change, it has been the guarantee that change would occur without too much upset' (Monteleone, 1992: 279-80). This balance of tradition and modernisation has been as important in the 1980s and 1990s as it was in the 1950s and 1960s. In the past, it served the interests of the ruling Christian Democrats. In more recent times, it has been skilfully manipulated by Berlusconi. Although Berlusconi has always taken care to present himself as a star figure, the incarnation of the dreams of the masses for comfort and glamour (Ortoleva, 1995: 62-65; Gundle, 1995a), the real key to his success first as a popular entrepreneur and then as a political leader lies in the extent to which he symbolises for Italians all the social, economic and cultural transformations of the last fifteen years. His television stations above all others remoulded Italian identity and encouraged segments of the population to embrace with enthusiasm the myths and values of a prosperous, dynamic society. It was on this basis, combined with Berlusconi's use of football and familiarity with retail and marketing strategies, that the Forza Italia movement was built.

Italian television remains traditional in a number of ways. In the first place, it is technologically backward. Although European co-operation grew in the 1980s, teletext services were developed and RAI has experimented with high-definition television, satellite broadcasting has not developed as an independent sector. Three channels covertly owned by Fininvest (Tele + 1, 2 and 3), whose ownership became the subject of judicial investigation in 1994, began broadcasting via the airwaves in the early 1990s in anticipation of the development of national cable television, but progress in this area has been slow. Moreover, despite debate about the fragmentation of mass audiences and the advent of 'narrow-casting', Italian television is still almost entirely governed by conceptions of quantity.

Secondly, political influence remains strong. The form of this has changed following the 1993 mini-reform of RAI, Berlusconi's entry into politics, and approval of 'fair play' rules during election campaigns. But, although party influence in RAI is now weaker than it was in the 1980s, depoliticisation has not been achieved.

Thirdly, traditional ideas about gender roles continue to hold sway. While women have in recent years occupied a few managerial roles (in public but not generally in private television) and are now employed as newscasters, journalists and presenters as well as announcers and soubrettes, there is still abundant use of women in a purely decorative capacity. Strong emphasis is placed on the exterior appearance of any woman who appears on screen. In this sense, the rapid liberalisation of mores has not significantly eroded a fundamental sexism in Italian society or created a genuine system of equal opportunity. As a male-dominated medium, television predominantly offers images of women that are attractive to men.

Finally, family and religion remain great reference points. Although church-going has declined massively since the 1950s and Italian families no longer share leisure habits (the multi-television household is a product of this), conventional pictures of social life are continually presented on screen in advertisements and family-oriented shows. Priests are frequent guests on programmes of all types and one or two have even acted as presenters. Shows such as *C'eravamo tanto amati* (We Loved Each Other Very Much), in which couples are encouraged to argue, and *Amici* (Friends), which aims to repair bridges in unhappy families, spectacularise conflicts, but also reinforce the image of the family as the central institution of society.

Bibliography

Bettinelli, E. (1995), *Par condicio: regole, opinioni, fatti* Turin, Einaudi.

Buttafava, G. (1980), 'Un sogno americano: quiz e riviste TV negli anni cinquanta' in Bettetini, G. (ed.) *American Way of Television: le origini della Tv in Italia*, Florence, Sansoni.

Bruzzone, M. (1981), 'Come la pubblicità ha influenzato l'evoluzione delle Tv private', *Problemi dell'informazione*, 6:4, 587-616.

Caretti, P. (1995), 'Le autorità independenti nel settore dell'informazione radiotelevisiva: alcuni modelli a confronto', *Problemi dell'informazione*, 20:3, 293-307.

Chiarenza, F. (1978), *Il cavallo morente: trent'anni di radiotelevisione italiana*, Milan, Bompiani.

Di Dario, V. (1992), *Pippo, Mike & Raffaella: la televisione italiana dagli esordi ai nostri giorni*, Milan, Sperling & Kupfer.

Eco, U. (1990), 'A Guide to the Neo-Television of the 1980s' in Baranski, Z. and Lumley, R. (eds.) *Culture and Conflict in Postwar Italy,* London, Macmillan.
Forgacs, D. (1990), *Italian Culture in the Industrial Era 1880-1980: Cultural Industries, Politics and the Public,* Manchester, Manchester University Press.
Giglioli, P. and Mazzoleni, G. (1990), 'Concentration trends in the media' in Sabetti, F. and Catanzaro, R. (eds.) *Italian Politics,* Vol.5, London, Pinter.
Grasso, A. (1992), *Storia della televisione italiana,* Milan, Garzanti.
Grasso, A. (1993), *Al Paese dei Berlusconi,* Milan, Garzanti.
Gundle, S. (1986), 'L'americanizzazione del quotidiano: televisione e consumismo nell'Italia degli anni Cinquanta', *Quaderni storici,* 21:62/2, 561-94.
Gundle, S. (1990), 'From Neorealism to Luci rosse: cinema, politics, society 1945-85' in Baranski and Lumley, *Culture and Conflict.*
Gundle, S. (1992), 'Italy' in Butler, D. and Ranney, A. (eds.) *Electioneering,* Oxford, Oxford University Press.
Gundle, S. (1995), *I comunisti italiani tra Hollywood e Mosca: la sfida della cultura di massa (1943-91),* Florence, Giunti.
Gundle, S. (1995a), 'Il sorriso di Berlusconi', *Altrochemestre* 3, 14-17.
Gundle, S. (1996), 'The Rise and Fall of Craxi's Socialist Party' in Gundle, S. and Parker, S. (eds.) *The New Italian Republic: from the Fall of the Berlin Wall to Berlusconi,* London, Routledge.
Gundle, S. (1996a), 'RAI and Fininvest in the Year of Berlusconi' in Katz, R. and Ignazi, P. (eds.) *Italian Politics in the Year of the Tycoon,* Oxford, Westview Press.
Gundle, S. (1996b), 'Fame, Fashion and Style: the Italian Star System' in Forgacs, D. and Lumley, R. (eds.) *Italian Cultural Studies: an Introduction,* Oxford, Oxford University Press.
Gundle, S. and O'Sullivan, N. (1995), 'The Crisis of 1992-1994 and the Reform of Italian Public Broadcasting', *Modern Italy* 1:1, 70-81.
Gundle, S. and O'Sullivan, N. (1996), 'The Mass Media and the Political Crisis' in Gundle and Parker, *The New Italian Republic.*
Martini, P. (1990), 'Molti affari, molta politica: nei rapporti con i partiti il lato forte di Berlusconi', *Problemi dell'informazione* 15:4, 513-28.
Mazzoleni, G. (1994), 'La Rai tra ristrutturazione e risanamento' in Mershon, C. and Pasquino, G. (eds.) *Politica in Italia: edizione 1994,* Bologna, Il Mulino.
McCarthy, P. (1996), 'Forza Italia: the New Politics and Old Values of a Changing Italy' in Gundle and Parker, *The New Italian Republic.*
Monteleone, F. (1992), *Storia della radio e della televisione in Italia,* Venice, Marsilio.
Ortoleva, P. (1995), *Vent'anni a colori: televisione privata e società in Italia (1975-95),* Florence, Giunti.
Pasquino, G. (1983), 'Mass media, partito di massa e trasformazioni della politica', *Il Mulino,* 288, 559-79.
Pettinati, S. (1988), *TIVU: cronaca della televisione,* Turin, SEI.
Richeri, G. (1990), 'Hard Times for Public Service Broadcasting: The RAI in the Age of Commercial Competition' in Baranski and Lumley, *Culture and Conflict.*
Veltroni, W. (1992), *I programmi che hanno cambiato l'Italia: quarant'anni di televisione,* Milan, Feltrinelli.
Wolf, M. (1989), 'Italy from Deregulation to a New Equilibrium' in G. Nowell-Smith (ed.) *The European Experience,* London, BFI Publishing.
Wolf, M. (1990), 'The Evolution of Television Language in Italy since Deregulation' in Baranski and Lumley, *Culture and Conflict.*

Television in Ireland

Eddie Holt and Helena Sheehan

History of broadcasting

Television in the Republic of Ireland was inaugurated on December 31, 1961. The new station, named Telefis Eireann, emerged after more than four years of debate and controversy over how it should be financed and structured. By the late 1950s, more than 30,000 homes were receiving the two existing British television channels, BBC and ITV. The ITV regional stations which could be received were Ulster Television (UTV) in the northern and north-eastern parts of the state and Welsh television (HTV) in the south-eastern counties.

Irish television viewers were therefore aware of the public service broadcasting model of the BBC and the commercial television model of ITV. Ireland's Broadcasting Authority Act, 1960, chose the public service model to be run under a mixed economy system of licence fee and advertising. But finance and structure were not the most contentious aspects of the TV debate. The role of television in traditionally conservative Ireland caused heated argument between isolationist conservatives and integrationist modernisers. It should be remembered that, in 1961, the Irish state had been in existence for just four decades and that the Irish Free State (which became the Republic of Ireland in 1949) had remained neutral throughout World War II.

Ireland, therefore, through not only geography, but through history, was politically and culturally isolated from the rest of Europe. The country was dependent economically on trade with Britain, but the dominant project of the young state was to assert its independence from Britain. In this climate of political and popular opinion, the Republic was an inward-looking state in which feudal ties of blood and land prevailed over the more modern European bourgeois ties of capital and labour. The economy was primarily agricultural and hierarchical authority - most significantly the authority of the Roman Catholic Church - dominated at the expense of democratic rights. Holy Ireland had endeavoured to keep itself to itself, erecting tariff barriers in the realm of economics and censorship laws in the realm of culture.

By the 1960s, however, it was clear that the outside world was encroaching. Emigration, capitalist market forces, European radio waves, British television, Hollywood films and imported books were making increasingly broad inroads. The modernisers, led by Sean Lemass, who had become Taoiseach (Prime Minister) in 1957, largely welcomed the opportunities offered by the incoming influences. The conservatives, led by the ageing President Eamon de Valera, feared these alien influences and a coalition of like-minded politicians, priests and patriarchs railed against the coming changes.

It was against this backdrop that on New Year's Eve, 1961, Telefis Eireann was launched. Beginning with the national anthem, *Amhran na bhFiann* (The Soldier's Song) the inaugural night's programming included music, poetry, news, benediction and ceremonial speeches, including a cautionary one from Cardinal D'Alton, the Catholic Primate of All-Ireland. But it was De Valera who rung the most cautionary note, likening television to atomic energy: it could be used either for good or bad. It could either impart knowledge and build the character of a whole people or it could lead 'through demoralisation to decadence and dissolution'. Lemass, in contrast, was full of optimism about what television could bring. He addressed the Irish people as citizens of the world as well as of Ireland, stressing that there were standards, aims and values that transcended national frontiers and were universal in application.

This debate between the conservatives and the modernisers has been ongoing since the inaugural night's programming. Both sides have recognised the power of television and the frequent and bitter ideological clashes, part of what has been termed Ireland's 'moral civil war', are, generally, more vividly remembered than such technological advances as colour transmission (introduced in 1971), the growth of cable and satellite channels and the ubiquity of VCRs. Inevitably, the forces of modernity have gradually overcome those of traditionalism, but ideological battles continue to rage sporadically on RTE. (In 1966, Telefis Eireann was renamed Radio Telefis Eireann - RTE.)

A key date in RTE's early history was the 50th anniversary of the 1916 Easter Rising, which had led to 26 of Ireland's 32 counties gaining independence from Britain. For those 1966 commemorations, the Republic devised lavish, militaristic pageants and its five-year-old television station ran a nightly series, *Insurrection*, which dramatised the events in Dublin of Easter week a half century earlier. Within three years, however, Northern Ireland, a separate state which remained part of the United Kingdom when Ireland was partitioned in December, 1920, was on the brink of civil war. Television images of black civil rights campaigners in the US had prompted the Northern Ireland Civil Rights Association (NICRA) to use the same tactics, and large scale street violence erupted in Derry and Belfast in August, 1969. From then on, RTE's most contentious battles, within and without, would be political. In 1972, the Minister for Posts and Telegraphs, Gerry Collins, sacked the entire RTE Authority (see under Key Legislation) over the broadcasting of a report of an interview with a spokesman for the Provisional IRA. Four years later, the Broadcasting (Amendment) Act 1976, inspired by the then minister, Conor Cruise O'Brien, included the highly controversial Section 31, which prohibited interviews with members of such named organisations as Sinn Fein, the IRA (Irish Republican Army) and their paramilitary enemies, the unionist UVF (Ulster Volunteer Force). The Section 31 order remained in force, passed by successive governments, until 1994 when the current minister for Arts, Culture and the Gaeltacht, Michael D. Higgins, revoked it as part of the preparation for the peace process in Northern Ireland.

In 1978, RTE launched a second television channel - RTE 2 (renamed Network 2 in 1988), by which time BBC 2, as well as BBC 1 and ITV, were available in most parts of the Republic. In 1982 Britain's Channel 4 was launched and it too could be received in most parts of Ireland. The 26-county Republic therefore had access to six terrestrial television channels. A seventh, Teilifis na Gaeilge (an Irish language channel) and an eighth, TV3 (an Irish commercial channel) are due to begin transmitting in late 1996 and mid 1997 respectively. Satellite and cable channels mushroomed throughout the 1980s and 1990s. Of these, Sky News, Sky Sports and Sky Movies have proven to be the most popular. (A comprehensive listing of satellite channels in Ireland is included under Available Broadcast Channels. See List of Key Dates.)

Key legislation, regulatory frameworks and links with Government

Under the terms of the Broadcasting Act of 1960, the responsibility for all broadcasting in the Republic of Ireland was vested in a statutory, independent authority (originally the Radio Eireann Authority, but since 1966, the Radio Telefis Eireann Authority). In the main, the model used was that of the BBC, with much stress being placed on the independence of the Authority from both the Government and the civil service (which had, in effect, controlled radio broadcasting since its inauguration in 1926).

In spite of this stress on the 'independence' of the Authority, the Act reserved crucial powers to the Minister for Posts and Telegraphs. These powers included the right to appoint or remove from office the members of the Authority and the right to issue directives in regard to broadcasting. Under the 1960 Act, the Authority was obliged, in performing its functions, to bear in mind certain 'national aims'. These were defined, specifically, as 'the restoration of the Irish language' and, more ambiguously, as 'the preservation and development of the national culture'.

Throughout its first decade and into the mid-1970s, RTE's relationship with Government was stormy. In a debate in the Dail (Ireland's parliament) in 1966, the Taoiseach Sean Lemass stated: 'Radio Telifis Eireann was set up by legislation as an instrument of public policy and, as such, is responsible to the Government ...The Government rejects the view that Radio Telifis Eireann should be, either generally, or in regard to its current affairs and news programmes, completely independent of government supervision.'

So, the *realpolitik* of the position of the nine-person RTE Authority was established: it was and continues to be a mediator between the broadcasters and the Government. The following year, a government directive prohibited an RTE news team from going to North Vietnam and in 1968 from visiting Biafra. But it was in 1972, as noted earlier, that the full power of the Government in relation to the Authority was exercised, when the Minister of the day, Gerry Collins sacked the entire Authority over the broadcasting of a report of an interview with a Provisional IRA spokesman.

In 1976, the Broadcasting Authority (Amendment) Act moved to defuse the rancour over the issue of independence, but introduced too - under its notorious Section 31 - the most contentious piece of censorship in the history of Irish broadcasting. Ministerial powers over the Authority were restricted to the extent that appointments and removals from office had to have the assent of both Houses of the Irish parliament (the Dail and the Seanad). But Section 31 of the Act, in its banning from the airwaves of certain political and paramilitary organisations, effectively castrated RTE on the most sensitive and dangerous running news story in Ireland - the conflict in Northern Ireland.

To deal with a growing number of 'pirate' radio (and, briefly, one television) stations in the 1980s, the Independent Radio and Television Commission (IRTC) was established in October, 1988. It is empowered to select and regulate organisations other than RTE in relation to the provision of certain broadcasting services. To date, almost all of its energies have been expended on independent radio, but TV3 - the Republic's first fully commercial television channel - is scheduled to begin transmitting in 1997, most likely in the autumn of that year. Teilifis na Gaeilge, intended to save and promote Gaelic, Ireland's ailing first national language, is due to start later in 1996, but it is, at present, the subject of much rancorous debate, particularly in regard to its financing, estimated at about

£15 million a year. Those in favour of this new Irish language channel argue that it is culturally necessary. Those against, insist that its cost will seriously undermine the already beleagured RTE and lead to a lowering of standards across the board in Irish television.

Funding and advertising

RTE is funded by a combination of annual licence fee, advertising and profits from ancillary companies. In 1995, licence fees brought in IR£51.1 million and all other broadcast revenue raised IR£83 million, which gave a total of IR£134.1 million. A decade ago, in 1986, licence fee income accounted for IR£38.1 million and other income for IR£44.6 million, giving a total of IR£82.7 million.

It is notable that in this past decade licence fee revenue, as a proportion of total income, has been declining. In 1986, it accounted for 46%. By 1995, it had slipped to 38%. This trend reflected the prevailing political ideology of the period, which, though less vehemently than in Britain, championed private capital over state enterprise. Repeated requests to Government from the RTE Authority to permit an increase in the licence fee, which between 1986 and 1994 had not been sanctioned (£62 for a colour set and £44 for a black and white set in 1995) were rejected. However, in 1996, an increase in the licence fees to £70 for a colour set and £52 for a black and white have been sanctioned. The Authority had sought a raising of the crucial colour fee to £75. (Though RTE denies it, it is the opinion of many critics that the station, particularly in relation to current affairs programming, became more timid as its funding grew increasingly dependent upon advertising revenue. Certainly, the figures lend financial credence to this observation and suggest that successive governments, after aggressive scrutiny in the late 1970s and early 1980s, were not displeased to see the station nudged towards a more commercial ethos.)

No less than programme content, advertising on RTE was contentious from the outset. Under the Broadcasting Authority Act, 1960, no political or religious advertisements were permitted and this ruling pertains to the present. The objection to religious advertising was based on the grounds that any religious group, including those which, in the view of the Authority, the majority of viewers might consider 'very objectionable and offensive' would have to be allowed to advertise. The objections against political ads were based on the principle that the national broadcasting service should not be used to further the interests of one political party over another. Less wealthy parties would, clearly, suffer if political advertising were permitted. Advertising relating to industrial disputes is also banned.

Television advertising on RTE is restricted to 10% of total broadcast time, or, as it is more commonly expressed, an average of six minutes in every hour. During peak viewing times - from 6pm to midnight - RTE may screen up to 7.5 minutes of advertising per hour, provided that the daily average does not exceed the stipulated 10% of total broadcast time. The new independent channel, TV3, because it will not gain revenue from licence fees, will be allowed to use 15% of its total airtime for selling advertising. With nine minutes advertising for every hour of broadcasting, the coming commercial channel will, not surprisingly, appear more like a US channel than an indigenous Irish one. Teilifis na Gaeilge, being state supported, will operate under guidelines similar to those regulating RTE.

Statistics and audience share

The Republic of Ireland has a population of 3.62 million (1996 Census) and more than one million television sets. (In 1995, more than 952,000 TV licences were sold. More than one in 20 homes now have more than one television, so the total number exceeds one million.) In 1961, the year that Irish television was inaugurated, the population was 2.81 million (1961 Census). There has, therefore, been a 35% increase in population in the Republic of Ireland in 35 years, a growth rate which is now steadily declining, but still leaves the state with the youngest (on average) population in western Europe. The growth in the number of TV sets was, not surprisingly, most dramatic during the 1960s and 1970s. By 1966, c380,000 homes out of a total of a little more than 700,000 were receiving RTE; by 1971, 536,000 homes had television. By 1984, sales of TV licences had reached 720,000 and in 1991, the 900,000 barrier was broken for the first time. Since then, the rate of increase has slowed and market penetration has almost reached saturation point. More than 97% of all homes in the state now have at least one television set.

Officially, cable television began in 1974 (although parts of Dublin and some of the other major cities and towns were cabled a little earlier). By 1995, there were 412,000 known cable subscribers. This figure represents c98% of the total number as the small, local suppliers, who originally entered the market, have been all but eliminated. In areas not suitable for cabling, Multi-Point Microwave Distribution Systems (MMDS) have become increasingly popular. By 1996, MMDS, which began in 1989 after legislation allowed it, had 76,500 subscribers. Satellite dishes, unlike the aerials of the 1960s, are far from ubiquitous. But as cable and MMDS companies offer the most popular satellite channels, the main period of growth for satellite TV is yet to come.

The introduction of satellite television channels in the middle to late 1980s (largely through cable systems) and the sale of satellite dishes in the early 1990s gave a fresh impetus to the expansion of channel choice in television homes. This situation is summarised in Table 2.

It is clear that the television environment has changed quite radically in Ireland over the past 20 years. The competitive challenges facing RTE have increased steadily with the expansion of choice for an increasing number of viewers, from

Years	**1995**	**1990**	**1985**	**1980**	**1975**
TV Homes	97%	94%	91%	91%	81%
Colour Sets	97%	91%	76%	58%	c30%
Multi-channel	69%	66%	55%	47%	43%
Dual Channel*	31%	34%	45%	53%	57%

*Dual Channel means homes receiving just RTE 1 and Network 2. These are largely confined to remote areas of the rural south and west of the country.

Table 1: TV Penetration and Multi-Channel Choice - % of All Homes

Years	**1995**	**1990**	**1985**
Cabled Homes	38%	40%	33%
MMDS Homes	4%	N/A*	N/A
Satellite Dish Homes	5%	1%	N/A
Satellite Receiving Homes	47%	35%	N/A

*N/A = Not Applicable

Table 2: Satellite Channel Reception in TV Homes 1985 to 1995

both terrestrial and particularly satellite channels, in the past decade. Ireland is unique in Europe in being open to a very large number of challenges from terrestrial and satellite channels broadcasting in the English language. Other small European countries may also have a large number of foreign terrestrial and satellite channels available through cable systems. However, the language factor operates as a barrier for a considerable percentage of the viewing public in Europe and helps to blunt some of the competitive challenge to domestic broadcasters.

In spite of the foreign challenge, RTE continues to dominate market share. Core programmes, which address and reflect Irish society on issues and in tones which foreign channels can not and the fact that almost one-third of Irish homes still receive only RTE 1 and Network 2, are principally responsible for this. Between them, in 1995, the two RTE channels had 60% of market share (42% for RTE 1 and 18% for Network 2) in the evening and night hours (6pm to midnight). The two BBC channels recorded a combined total of 15% (BBC 1 had 11% and BBC2 4%). Britain's independent commercial channels won 18% of viewers (ITV's regional channels, UTV and HTV had 13%; Channel 4 had 5%). The combined satellite channels commanded the remaining 7% of market share. The less lucrative 24-hour figures showed RTE with 54% of the total; British terrestrial channels with 36% and the satellites with 10%.

Description of individual channels

RTE's two television channels - RTE 1 and Network 2 - broadly replicated, until recently, the mainstream/minority division of the UK's BBC 1 and BBC 2. There were significant differences, of course, mostly caused by the huge gap in the scale of resources netween RTE and the BBC. RTE 1's programme formats have reproduced those developed by American network television and BBC 1: news, current affairs, talk shows, drama serials, quiz shows, arts shows, sports coverage and documentaries. Network 2, though, relied much more heavily than BBC 2 on imported programmes. In the last couple of years, however, Network 2 has increasingly been dominated by sports coverage (all but excluded now from RTE 1 schedules). The growing audience for televised sport and the success of Sky Sports have been responsible for this fundamental change in Network 2's schedules. The fact too, that sports programming, while requiring technical expertise, is essentially about coverage rather than production, makes sport financially appealing to RTE.

Talk shows are the dominant genre of RTE's home-produced programming. There are two principal reasons for this: (i) talk TV, being studio-bound, is inexpensive; (ii) television in the Republic of Ireland, as noted, has been characterised by facilitating public debate in a country which, socially, has changed more dramatically than any of its western European neighbours in the last three and a half decades. It would be impossible to write a meaningful account of Irish television without considering *The Late Late Show*, the longest-running (since 1962) chat show in the world and the country's most influential programme for more than 30 years. Presented by Gay Byrne, easily the most recognised person in Ireland, the show has been credited with playing a crucial role in the liberalisation of Irish society. This, up to a point, is true, but Byrne's show has effected change because he has acted as a sensitive barometer of what society could bear - of what currents could no longer be contained - rather than as a conscious and committed initiator of social change.

But there is a less meritorious aspect to this concentration on talk TV, often disparagingly dubbed 'radio with pictures'. In attempting to economise and to bandwagon on the success of *The Late Late Show*, RTE has overused the format with the inevitable result that other genres have been neglected. Drama, a staple in the early years, has declined noticeably and the single play has virtually disappeared. After a rich history of a high volume of studio-based single plays in the 60s, to more on-location series and mini-series in the 70s, to more expensive co-productions (frequently with Britain's Channel 4) in the 80s, virtually all that is left in the 90s are two soap operas, *Glenroe* (rural) and *Fair City* (urban). RTE's drift from a public service ethos to ratings-driven commercialism is, arguably, as clearly illustrated in its diminishing drama output as in its increasingly timid and restricted current affairs programming.

Still, in spite of RTE's loss of early enthusiasm and the drift towards more commercial broadcasting, television culture in Ireland is extremely cosmopolitan. Indeed, it can be argued that Irish viewers occupy a privileged position with respect to exposure to the best of the world's television. With British television - terrestrial and satellite - now available in most parts of the island of Ireland, viewers enjoy great choice. A little less than 50% of RTE's 10,500 hours of yearly television is home produced. The imported material comes, not unexpectedly, from the US but the sources for imported programming have expanded considerably in the past decade. The tolerance of English-speaking audiences - and Ireland is no exception - for subtitled or dubbed material is still comparatively low, but the persistence of a public service ethos in European broadcasting assures such material an airing for the specialised audience who want it, despite low ratings. By far, however, the largest amount of imported material consists of US prime-time series, mini-series and TV movies. Australian soap operas have consolidated their places in RTE's schedules since the early 1980s.

It is important to stress that RTE has never had a monopoly on its audience with the result that indigenous Irish programming has always developed in the context of its exposure to international television, both on the parts of its audience and its producers. Because of this exposure, its production standards have tended to be relatively high, often prompted by fear of comparisons with the BBC. Nonetheless, Irish television has its distinctiveness, sometimes in pioneering formats, but more importantly in its positioning in relation to the community whose collective consciousness it helps to create and reflect. The ratings successes of programmes such as *The Late Late Show*, *Kenny Live* (another prime-time talk-show), *Glenroe* and *The Riordans* (a pioneering rural soap which ran from 1965 to 1979) against all foreign competitors, including such American 'world hits' as *Dallas*, *Hill Street Blues*, *Cheers* and Britain's most popular TV programme, the soap *Coronation Street*, proves the point.

In its Mission Statement, RTE defines its role as ' . . . to ensure that its two channels continue to be the preferred choice of viewers on this island and to do so by transmitting a range of quality home-produced programmes complemented by the best of international television. . . . The Television Programmes Division, which is responsible for all television output except News, employs approximately 360 people, most of whom are engaged in programme-making . . . Current expenditure on programming amounts to IR£42 million, with an additional IR£17 million being incurred in Television Production Facilities such as sound, lighting, studio maintenance and editing facilities. Programme production is organised in areas of specialism - Sport, Current Affairs, Features, Young Peoples and Entertainment - and the overall management structure is designed to facilitate maximum efficiency within an environment of clearly defined delegated responsibilities.... RTE intends that its Television services will remain the preferred choice of its viewers, retain its national audience share of 60% and increase its share in multichannel areas, by enhancing its schedule, starting transmission earlier in the morning and continuing later at night. . . . Redevelopment of new indigenous programmes is a high priority with a particular focus on the areas of comedy and drama. Priority will also be given to the areas of Pre-School, Children's and Young People's programmes; Irish language programmes and current affairs; to the stimulation of the independent production sector and to the increase of regionally sourced and produced material.'

Satellite and cable channels

As shown in Statistics and Audience Share section, most satellite channels are received through cable and MMDS systems. Almost half (47%) of all Irish TV homes receive one or more satellite channels despite the fact that only one in 20 such homes has a satellite dish. (See Table 1).

The satellite channels available can be categorised under the headings of News, Sport, Movies, General, Music and Children's, although foreign language and the euphemistically titled 'adult' (i.e. porn) channels can also be accessed by the use of viewing cards. News, sport and movie channels are the most popular, in particular, Sky News, Sky Sports (because it has exclusive rights to screen live English Premiership soccer, which, having many Irish players, is avidly followed in Ireland) and Sky Movies. For the most part, however, local cabling arrangements and cost (most satellite channels are available on a subscription basis) determine the selection and spread of such channels in domestic homes.

Sports channels available include Sky Sports, Sky Sports 2, Sky Sports Gold and Eurosport. Movie channels include Sky Movies, Sky - The Movie Channel and Sky Movies Gold. It is clear that Rupert Murdoch's Sky channels dominate to a degree approaching monopoly in news, movies and sport. General channels, either because their schedules are jammed with repeats or are aimed at very specific audiences are not as much in demand. But there is a wide range available. These include NBC Super Channel, Sky One, UK Gold, Bravo, The Discovery Channel, The History Channel, Sky Soap, The Disney Channel, UK Living, The Family Channel, TLC (The Learning Channel), Sci-Fi Channel, Travel TV, Sky Travel and an arts channel, Performance. Among music channels, the most popular, by far, is MTV, but VH-! (another pop and rock channel) and CMT (Country Music Channel) are also available. Children's channels include TCC (The Children's Channel), Cartoon Network and Nickelodeon. Between them all, however, it must be remembered that in 1995, they had just 10% of market share and just 7% between 6pm and midnight. Their real impact has yet to come.

Key Dates

1961	Telefis Eireann (since 1966 RTE) is launched.
1962	*The Late Late Show*, still running (see text) begins.
1965	*The Riordans*, a rural soap opera, which regularly laced its plots with the burning social issues of the time, begins. It ran until 1979.
1966	*Insurrection* commemorates the 50th anniversary of Dublin's Easter Rising which led to independence from Britain for 26 of Ireland's 32 counties in 1921. Taoiseach (Prime Minister) Sean Lemass asserts that RTE 'is responsible to the Government'.
1970	Ireland wins the Eurovision Song Contest for the first time. Six further wins (in 1980, 1987, 1992, 1993, 1994 and 1996) make Ireland the most successful country in the contest. But staging Eurovision so frequently places severe strain on RTE budgets.
1971	RTE begins transmitting in colour.
1972	The RTE Authority is sacked by the Minister for Posts and Telegraphs.
1976	The Broadcast (Amendment) Act, 1976 (Section 31) bans interviews with selected (see text) paramilitary organisations and their political supporters. It is censorship which seriously distorts reporting on the conflict in Northern Ireland.
1978	RTE 2 (renamed Network 2 in 1988) is launched. A drama serial, *The Spike*, is axed in mid-run because of a nude scene.
1980	*Strumpet City*, RTE's largest ever drama production, costing £1 million, is screened.
1982	Britain launches Channel 4 and an era of RTE-Channel 4 drama co-productions begins.
1985	Independent consultants Stokes, Kennedy, Crowley (SKC) recommend 320 staff cuts and a 35% increase in home production if RTE is to stay competitive.
1988	The Independent Radio and Television Commission (IRTC) is formed.
1990	A cap, designed to aid commercial broadcasters, is placed on RTE advertising. It was lifted in 1992.
1994	Section 31 is revoked.
1996	Teilifis na Gaeilge (an Irish language channel) due to begin.
1997	TV3 (the Republic of Ireland's first fully commercial TV channel) due to start.

Bibliography

Books

Bew, Paul and Patterson, Henry (1982), *Sean Lemass and the Making of Modern Ireland 1945-1966*, Dublin, Gill and Macmillan.

Campbell, James (1961), *Television in Ireland*, Dublin, Gill and Son.

Doolan, Lelia, Dowling, Jack and Quinn, Bob (1969), *Sit Down and Be Counted*, Dublin, Wellington Publishers.

Farrell, Brian (ed.) (1984) *Communications and Community in Ireland*, Cork, Mercier Press.

Fisher, Desmond (1978), *Broadcasting in Ireland*, London, Routledge and Kegan Paul.

Gibbons, Luke (1988), 'From Megalith to Megastore: Broadcasting and Irish Culture' in Bartlett, Thomas, Curtin, Chris, O'Dwyer, Riana and O Tuathaigh, Gearoid (eds.), *Irish Studies -A General Introduction*, Dublin, Gill and Macmillan.

Gorham, Maurice, (1967), *40 Years of Irish Broadcasting*, Dublin, Talbot Press Limited.

Hall, Eamonn (1993), *The Electronic Age: Telecommunication in Ireland*, Dublin, Oak Tree Press.

Lee, J.J (1989). *Ireland 1912-1985: Politics and Society*, Cambridge University Press.

McLoone, Martin and MacMahon, John (eds.) (1984), *Television and Irish Society: 21 Years of Irish Television*, Dublin, RTE-IFI.

Savage, Robert (1996), *Irish Television: The Political and Social Origins*, Cork University Press.

Sheehan, Helena (1987), *Irish Television Drama: A Society and its Stories*, Dublin, RTE.

The RTE Book (1989), Dublin, Town House.

Tydeman, John and Kelm, Ellen Jakes (1986), *New Media in Europe: Satellites, Cable, VCRs and Videotex*, London, McGraw-Hill.

Periodicals and Reports

Bell, Desmond (1985), 'Proclaiming the Republic: Broadcasting Policy and the Corporation State in Ireland' in *Broadcasting and Politics in Western Europe*, 8, no. 2.

Feeney, Peter (1984), 'Censorship and RTE' in *The Crane Bag*, Dublin, vol. 8, no.2.

Independent Radio and Television Commission, *Codes of Standards, Practice and Prohibitions in Advertising and Other Forms of Commercial Promotion in Broadcasting Services*, Dublin.

Irish Broadcasting Review (1978-1983), Dublin.

Irish Government's Green Paper on Broadcasting (1995).

Kearney, Richard. (ed.) (1984), 'A Debate on the Media and Popular Culture' in *The Crane Bag*, Dublin, vol.8, no. 2.

RTE (Radio Telefis Eireann) annual reports.

RTE Guide (weekly listings and features television magazine).

RTE (1989), *Broadcasting Guidelines for RTE Personnel*, Dublin, RTE.

RTE's Response to the Green Paper on Broadcasting (1996).

Sweeney, Maxwell (1963). 'Irish Television, a Compromise with Commerce' in *Studies*, Dublin, 52.

Newspapers

Irish Independent,
The Irish Times,
The Sunday Business Post,
The Sunday Independent,
The Sunday Tribune,
The Examiner.

(Also, *Irish Press* and *Sunday Press*, though defunct since 1995).

Television in Belgium

Daniël Biltereyst

Introduction

In international literature on television, Belgium is often quoted as an interesting but complex case. The Belgian broadcasting system is mostly regarded as a very open one with the highest national cable penetration rate in the world (95.2%), but also as one whose organisation and policies are strongly determined by internal cultural and linguistic diversities. In this context it is safe to say that the history and structure of Belgian television reflect the slow but resolute process of political federalisation. Since the new Constitution (17 February 1994), Belgium is regulated as a federal state with three main language communities: a Flemish Dutch-speaking community (of some 5.7 million people), a French-speaking community (4.2 million) and a marginal German-speaking community (65,000 people). The transfer of powers from the central to the regional level however has a longer history, especially in respect of cultural matters including broadcasting. One can even say that since the 1960s there has been a tangible disintegration of Flemish- and French-speaking broadcasting, giving rise to divergent and autonomous regional television activities, policies and ethos in the northern (Flanders) and southern part of the country (Wallonia).

The Belgian broadcasting system may be a complex one due to the different internal cultural and broadcasting policies, but it is surely not a-typical from a European viewpoint. On the contrary, as a small country in the heart of the European Union (EU), Belgium has witnessed since the 1980s most of the main trends in European television, such as the process of demonopolisation and deregulation; the increasing number of local stations and the extension of television output; the advent of new private competitors; the financial, organisational and programming crisis of the local public service broadcasting (PSB) stations; or the growing power of transnational advertising and multimedia actors. In this article we first present an historical background in order to understand the present-day Belgian television scene. Then we focus on the Dutch- and French-language television landscapes separately, given their strong divergences.[1]

Historical overview: from public service monopoly in the old nation state to a federalised multi-channel environment

Comfortable public service monopoly (1930 - 1977)

Until the second half of the 1980s, Belgian television was dominated by a tradition of strong public service monopoly. The creation and reinforcement of this PSB-monopoly went through different stages. In 1930 the existing commercial radio broadcasting scene was replaced by a single national public broadcasting corporation, the NIR/INR - which was a compromise between a state broadcasting monopoly (the British tradition) and a pillared system with independent broadcasting associations of ideological and political descent (the Dutch tradition). After the traumatic (broadcasting) events of the second World War, the NIR/INR-monopoly was re-established and even strengthened, mainly by abandoning the private associations. This monopolistic corporation soon became strongly controlled by the political parties in power, mainly Catholics, Socialists and Liberals.

1. In order not to complicate the picture of Belgian Broadcasting, we will mostly concentrate on the major national and broad-appeal television stations designed for a general audience. This means that we do not elaborate on the structure and policy of local and community channels, or of marginal pay-TV stations.

It was this public broadcasting corporation which was soon commissioned to introduce television in collaboration with local electronic industries. Compared to its neighbouring countries such as France (1945), Germany (1950) and the Netherlands (1951), regular television services started fairly late in Belgium (31 October 1953). This was mainly due to technical problems, related to the fact that television sets needed to be developed which could receive both the international standard of 625 lines and the French one with 819 lines. In fact, during the late 1940s there was a heavy dispute over television standards (the 'lines battle') in the still unitary Belgium, where the Dutch-speaking part opted for the international standard, while the Walloons chose the higher screen definition of the French system. Black and white television broadcasting finally started in 1953 and colour in 1971. The penetration of television was boosted by the 1958 World Exhibition in Brussels, but only became really widespread during the 1960s. According to official statistical records, it took 14 years for monochrome television to reach half of Belgian households. Although official data tend to be quite unreliable, the point of near-saturation (+ 95%) can be situated during the second half of the 1980s. In 1994 approximately one fifth of Belgian households possessed two or more television sets (OEA, 1996).

Besides the growing popularity of television, the 1960s have also been characterised by a further regionalisation of the Belgian broadcasting system. A decisive step in this direction was made by the 1960 broadcasting act, transforming the national NIR/INR into two autonomous corporations: a Flemish BRT and a Walloon RTB. In literature on Belgian broadcasting and cultural policy in general, the 1960 act is often seen as a fundamental legal restructuring of the broadcasting system, and even as an important cultural 'precursor' of a federalist policy, which will later on transform Belgian society into different communities. However, the new act hardly changed the relationship between political power and broadcasting. In fact, from the 1960's even into the 1990s, the Belgian public service broadcasters underwent a process of increasing bureaucratisation and politicisation. The latter can be seen at different levels, such as harsh party political control and pressure on the general management, the board of governors, the often politically appointed personnel, promotions, and sometimes even direct interference from political headquarters in journalistic work. The impact of all this on the internal/external functioning and the societal role of the Belgian PSB can not easily be underestimated, and it did not improve the public image and the general credibility of the monopolistic BRT and RTB in their respective communities.

Another important factor in the development of Belgian television in the 1960s and 1970s is the explosive cable penetration. In 1961 the first cable experiments were started up in the hilly southern part of the country, soon followed by local schemes all over the country. Especially during the 1970s, cable became widely promoted and supported by local authorities, often in collaboration with strong private electricity and energy lobbies. While in 1972, for example, 9.3% of Belgian television households had access to cable, ten years later a cable penetration rate of 87.1% was registered. This went up to 92% in 1990 and today cable has reached the point of near-saturation (95.2%). It is interesting to notice that, unlike most European countries, Belgium has thus witnessed a long tradition of openness and viewer choice: due to its internal cultural diversities, its geographical position and the high cable penetration rate, most Belgian households have, since the late 1970s, been able to choose between 15 or more channels.

Currently, the 34 cable operators serve some 3.5 million households, offering up to 30 channels. Mostly these are generalist channels from neighbouring countries, although some thematic channels (such as CNN, MTV) are often included in the cable package.

One important outcome of the early, wide penetration of cable, was that the local PSB-stations, even when they were considered to be monopoly service providers, were soon confronted with the phenomenon of competition. The further introduction of cable in Flanders, for example, went hand in hand with a quite dramatic decrease in the BRT market share during the 1970s and 1980s, given the fact that a significant proportion of the Flemish audience was strongly attracted by the more commercial programming style of some Dutch broadcasting organisations. A similar phenomenon can be observed in Wallonia, which has often been (and actually more than ever is) considered by French and Luxembourg-based television organisations (e.g. TF1 and CTL) as an attractive hunting ground for audiences and advertising budgets. This external competition and foreign challenge led the Belgian public broadcasters to introduce more popular programming categories during the 1970s. In order to diversify its output both public broadcasters reluctantly launched a second channel in 1977, designed for special events, sports and complementary programming.

Continuing process of demonopolisation and federalisation (1977 - 1987)

Another tendency of Belgian television in the 1970s was its definitive regionalisation, perfectly reflecting the general political process of federalisation of Belgian society: the broadcasting act of 18 February 1977 finally 'wiped out' the last traces of national television activities, which still existed mainly in the form of technical and administrative collaboration between the BRT and RTB.

In a process parallelled in most Western European countries, the 1980s were characterised in Belgium as a decade of discussion and policy-making directed towards the liberalisation and demonopolisation of television. As many authors have been pointing out in this respect (e.g. McQuail and Siune, 1986), these tendencies can best be interpreted as a result of a mixture of economic, political, socio-cultural and technological factors. Within Europe, however, quite different patterns can be observed in the complex process of transformation of the old monopolistic, nationally oriented and PSB-model of broadcasting into the new multi-channel, more commercial one. Concentrating on the Belgian case, it is important to note that given the internal divergences in organising broadcasting, the discourse and policy on deregulation took quite different forms in the two communities - although some parallels can be found.

This is surely the case for the technological factor. It is safe to say that during the 1970s and 1980s the general adoption of several technological 'innovations' such as cable, the remote control and VCR[2], revolutionised existing television practice and viewing habits. In Belgium one can not underestimate the impact of cable technology on the political-economic discourse in favour of breaking up the PSB-monopolies. The argument went that the high cable penetration gave rise to the success of certain foreign stations and the decreasing appeal of local broadcasters, which meant a de-facto demonopolisation of local PSB television. In 1988, for example, just before a commercial local station (VTM) was launched in Flanders, both BRT-channels attracted only some 57% of the local audience, while the Dutch stations had an average market share of 28%. In

In Belgium, the penetration of the VCR has been relatively slow compared to the US and the UK – which is mainly due to the tremendous success of cable. However, the growth of the VCR has been resolute from the second half of the 1980s onwards: the penetration went from 2.2% in 1980 to 43% in 1990. In 1994 a figure of 61% has been registered (OEA, 1996).

	1980	1990	1993
Television sets (*% of households*)	85.2	95.2	95.9
Number of black and white sets (*in millions*)	1.1	0.3	0.1
Number of colour sets (*millions*)	1.8	2.9	3.2
Total number of sets (*millions*)	2.9	3.2	3.3
Cable (*% of households*)	77.0	92.0	95.2
Satellite dishes (*millions*)	—	—	0.003
VCR (*% of households*)	2.2	43.0	61.0
Video rental (*number of cassettes in millions/year*)	7.4*	22.0	15.5
Video sell-through (*millions/year*)	0.4**	1.1	4.6
Cinema (*number of visits in millions/year*)	20.6	17.1	19.3

(*): data for 1984; (**): data for 1989.

Table 1: Key data on Belgian audio-visual equipment and expenditure

Wallonia a similar scenario could be observed for the RTB, with the Luxembourg-based CLT-channel RTL as the main challenger. One of the consequences of this situation was that Belgian industry had no direct access to the advertising-free local public stations and were obliged to advertise their products on foreign (Dutch or Luxembourg-based) channels. The argument that tremendous amounts of advertising money were flooding away, depriving local (audio-visual) industries, was often heard in political and economic debates on the introduction of commercial television in Belgium.

The most important argument for liberalising broadcasting must then be found in a mixture of political and economic reasoning. Of prime importance were the growing interests of local capital. In Flanders, the traditional local publishers were looking for new ways to diversify their activities, given the severe crisis in their core business (newspaper printing). In the French-speaking part of the country a similar movement could be observed, although the interest of some powerful financial lobbies was equally crucial. In particular, the Belgian holding Groupe Brussels-Lambert, with its major interests in the transnational broadcasting group CLT, tried to consolidate and to intensify its position on the local broadcasting scene.

All these economic forces perfectly matched the general policy aspirations of the conservative parties in power during most of the 1980s, which were those of liberalisation, stimulating private enterprise and the gradual withdrawal of the state. Translated into the Belgian broadcasting policy of the 1980s, this meant that the different governments, mostly of a liberal and Christian-democratic stamp, thought of television as an underdeveloped growth area; that it should be exploited by private local (print) media actors; and that the PSB-system needed to be revitalised and its monopoly broken up.

This broadcasting policy was also reinforced by the rather negative public image of the PSB-corporations, which were often seen as 'unapproachable', heavily politicised, bureaucratic and paternalistic state apparatuses.

Besides technological, political-economic and socio-cultural factors in liberalising Belgian television, one also needs to refer to a more politico-sociological reason related to the changing relationship between politics and broadcasting. In fact, over the late 1970s and the 1980s, observers witnessed a slow but determined process of decreasing party political control over the PSB-corporations. During that period a new, more critical generation of journalists - often labelled the post-May 68-generation - came to occupy ever more important positions within the radio and television editorial staff. Especially among many BRT-journalists, party loyalty became suspicious, while uncritical and distanced forms of journalism had to make room for more committed styles such as hard political interviews or investigative work. In daily reporting this new relationship between broadcasting and politics led to numerous confrontations, and eventually to the awareness that party political control over the public broadcasters (through its personnel, information, etc.) had very considerably diminished. This may sound like a *fait-divers* in the history of the demonopolisation and liberalisation of the Belgian television scene. However, the dominant (conservative) political establishment had found in it an important new motive for its broadcasting reforms: in the general debate Christian-democrats and liberals continuously claimed that the main information source in society (radio and television news) was monopolised by a 'red' staff, and that competition was needed to restore the (information) balance.

Overall, then, it is clear that a mixture of arguments have to be taken into account in understanding why television competition became the norm in Belgium in the 1980s. We could say that the arguments about a 'red' public broadcaster and the changed relationship between television and political headquarters were important, but may have been used as a legitimation for reform. The other financial and more political-economic factors may, however, have formed the real background for liberalising television.

Competition in a multi-channel environment (1987-)

From a legal perspective, 1987 may be considered as a milestone in Belgian television history. In Flanders, the cable decree of 28 January 1987 established the legal framework for starting up one commercial, non-public television station for the whole Dutch-speaking community in Belgium, for a period of 18 years. The decree also stated that this channel would have a monopoly over advertising, and that 51% of the shares would be held by local print media. Soon a consortium of local magazine and newspaper printers received the Flemish licence, and started with what would become one of the major success stories in European commercial television: VTM started in February 1989 and quickly became the market leader with more than 36% in 1990. The public broadcaster's market share (both channels combined) fell from 57% in 1988 to 29% in 1990. The arrival of the commercial VTM also triggered a decline in the appeal of the Dutch channels (from 25% in 1988 to 15% in 1990).

Besides establishing the legal framework for one commercial station for the whole Flemish community, the 1987 cable decree also regulated other television activities, such as pay-TV and local television. The former is exemplified by the rather unsuccessful pay-TV channel Filmnet (part of Multichoice), which has been active in Flanders since 1985, but which only became legalised with the new cable decree. The decree finally also established the legal framework for the start of one local television station in more than 10 regions within the Flemish community. During the period 1993-1995 up to eleven local stations have been launched in Flanders, with an average audience reach of 470,000 viewers.

While in Flanders the new cable decree legally transformed the broadcasting scene - with the start of ever more channels on different local and regional levels - a quite different situation can be found in the French-language community. First of all, local television channels had been allowed there since the 1970s. Today, Wallonia counts eleven local TV stations, with an average reach of 300,000 viewers.

Secondly, one could also speak of an unofficial commercial station for Wallonia, long before the new legal framework on the introduction of private television (decree of 17 July 1987): the Luxembourg-based station RTL, over 80% Belgian-owned, has been widely distributed on local cable since the beginning of the 1980s and was mainly directed to the Belgian audience and advertising market. The new 1987 decree on French-language commercial television might be seen as designed to give RTL a more legal and official Belgian profile. Soon after the new decree, RTL won the licence for 9 years, changed into RTL-TVi and became the official commercial channel of the French community. Unlike the Flemish VTM, however, RTL-TVi did not become a major success, mainly because of the strong competition from the privatised French channel TF1 which from 1987 succeeded in attracting a major audience in the Belgian French-language community (21.2% in 1990).

Finally, the 1987 decree on television in the French community also saw the launch of a pay TV station. The major French television company Canal Plus was given permission in 1988 to launch a local channel: the French parent controls 42.7% of Canal Plus Belgique, while smaller shares are held by local partners such as Benelux Pay TV (25%) and RTBF (10%). Like Filmnet in Flanders, Canal Plus Belgique cannot be called a major success, with a market share of 2.5%.

It is safe to consider the 1980s as the pivotal decade in both Belgian communities in legally transforming the monopolistic PSB-oriented television system into a more commercial and competitive one. But it was only at the end

National (federal) level	
June 1930 (law)	creation of the PSB-corporation NIR/INR
October 1953	start of regular television services
May 1960 (law)	NIR/INR changed into BRT and RTB
1971	colour TV
February 1977 (law)	final regionalisation of public television
French-speaking community	
December 1977	decree on the status of RTBF
July 1987	decree on broadcasting, giving way to commercial television
July 1991	legal framework for broadcasting throughout the French-speaking community
Dutch-speaking community	
December 1979	decree fixing BRT statutes, organisation and mission
January 1987	cable decree
February 1989	VTM launched and public service monopoly broken
June 1994	new cable decree on broadcasting
February 1995	start of VT4 (SBS) and Ka2 (VTM)
December 1995	mini-decree on the transformation of the BRTN

Table 2: Key dates in Belgian television history

of the decade, especially during the first half of the 1990s, that the tremendous consequences of this 'paradigmatic' change in the television industry became clear. On the positive side, the new situation led to a welcome boom in local audio-visual activity - such as the creation of new jobs and the growing importance of the independent video production sector; the increase and the redirection of advertising spending in favour of commercial television; or the arrival of ever more channels and the extension of television output, increasing viewers' choice. On the other side, and as in most Western European countries, one can say that the most visible consequences of the 'paradigmatic' shift were all in one way or another related to the heightened competition between the various stations. Here we can refer to such diverse aspects as the lost advertising revenues for traditional print media; the more audience-oriented or entertainment-based programming styles; or the heightened competition and rising costs involved in programme purchase, local popular programming and personnel.

All these diverse consequences have been still more intensified since the arrival of a second wave of new commercial stations in 1995. In that year, in both parts of the country, new commercial challengers entered the battlefield. The most striking new operator in the Flemish television scene has been VT4 (Viking Television), launched by the American-Scandinavian SBS (Scandinavian Broadcasting System, with ABC-Capital Cities as the most significant shareholder), which already operated channels in Denmark, Norway and Sweden. Although VT4 is officially registered as a British non-domestic licensed station (headquarters in London), it can be considered as a 'Flemish' station targeting the local audience and advertising budget, working with local personnel, management and production centres, and scheduling a mixture of local and American programming. The arrival of VT4 not only created a new and serious challenge to the existing duopoly and to VTM's advertising revenues, but also put the Flemish government and its broadcasting policy into disarray. In particular, the fact that the 1987 cable decree gave VTM, as the only commercial press-backed station in Flanders, an eighteen-year monopoly over television advertising, was interpreted as going against the grain of the European Television without Frontiers Directive. In June 1994 the Flemish government did adopt a new cable decree, in which certain protectionist articles of the 1987 decree were altered according to the EU-Directive, but in which it still tried to maintain its control over those (foreign) broadcasters wanting to enter the cable system. After several months of juridical and political fencing at the end of 1994, the Flemish government did not succeed in keeping VT4 off local cable, and in February 1995 the new commercial station started its conquest for a so-called 'unserved audience' (mainly young, educated men).

In order to defend its dominant audience and advertising market share against this Viking invasion, the leading commercial broadcaster VTM asked for and quickly received a new domestic licence in order to launch a second channel: Ka2 started two days before VT4. In the same month (15 February), in Belgium's French community, a new private television broadcaster was launched: CLT started, through its Belgian subsidiary RTL-TVi, with the dedicated movie channel Club RTL.

None of these new 'small' channels became a big success, with market shares ranging between two and five per cent, and they all changed their programming schedule within a couple of months. But the further fragmentation and competition on the local television market did change the relationship between television channels and the advertising sector, with lower tariffs and decreasing revenues for major commercial channels such as VTM.

Still, the most vulnerable victim of the new multi-channel environment in both Belgian communities has been the local public broadcaster. Since the end of the 1980s both the RTBF and the BRTN have been struggling with the question how to survive financially and how to live up to their traditional PSB-mission. One of the most important strategies in the most recent period of Belgian television policy, has been the search for a new equilibrium between public and private broadcasters. It may in this context sound quite strange that after a period of strong liberalisation on a regional, national and European level, the governments in both Flanders and Wallonia have been looking for a '*pax media*' (media peace) and ways of ordering the broadcasting scene. The most difficult task in this respect has been to restore the financial strength and the societal role of the local PSB-stations. Since the end of the 1980s, in both communities, several major legal initiatives have been launched in this direction - mostly including a new management structure, a new mission, and a more diverse, stronger financial backing. In all cases one can speculate about the growing commercial ethos within the Belgian PSB-stations - with the introduction of more corporate sponsoring and advertising on television, led by new (politically) independent managers with a private enterprise background. In both the BRTN and RTBF a greater emphasis has been laid, in recent years, on increasing audience shares, with a programming schedule which tries to blend the original PSB-mission and the strategy of imitating commercial stations. Given the differences between the communities, we will elaborate on these tendencies in the following sections, concentrating on the power relations and the new programming strategies within the Flemish and French-language television scene.

Sputtering cultural protectionism in Flanders

With the arrival of VT4 and Ka2 in February 1995, the number of indigenous television channels suddenly rose from three to five: two advertising-free BRTN stations, two VTM channels (VTM and Ka2), and one foreign station targeting Flanders (VT4). Many local media observers speculated about whether the small Flemish market can bear five generalist stations, with three of them living on advertising revenues. Looking at the evolution in the audience shares of the most important channels on the Flemish television market (Table 3), one can see that the introduction of new channels first of all went to the detriment of the three Dutch PSB-channels located on Flemish cable: while they still held 17.1% in 1989, their market share has been melting away to less than 5% in 1996.

A second observation is that the growing number of stations did not weaken the dominant position of VTM's first channel. On the contrary, the market leader on the Flemish television scene succeeded in strengthening its share during the tumultuous period, going from 36.4% the year before the introduction of VT4

	1989	**1990**	**1995**	**1996** (May)
BRTN TV1	22.5	21.8	17.5	21
BRTN TV2	9	7	5.1	4.3
VTM	27.4	35.6	37.1	41.1
VTM Kanaal 2	—	—	3.9	7.1
SBS VT4	—	—	6.1	7.7
Dutch Channels (n=3)	17.1	11.5	6.4	4.6
Others	24	24.1	23.9	14.2

Table 3: Market shares of the main channels in Flanders

and Ka2 (1994) up to 41.1% in 1996. The reason behind this unexpected success was that VTM's clean profile as the reliable family channel more than ever enjoyed the loyalty of the traditional viewing audience, and that it sharply contrasted with the new channels. But a key commercial factor has been VTM's ability to exploit the audience's need for cultural proximity: VTM's trump card in this respect has been its massive production of local drama series, using and transforming proven transnational programme formulas (sitcoms and soaps) into local, easily recognisable and popular products. During the period 1990-93, for example, VTM's local drama output grew from 28 to 226 hours. In order to defend its market share, the programming department even increased the frequency of this type of very expensive, but highly popular local fiction. In general, VTM's programming strategy is strongly oriented towards fiction in general (up to half the global output) and entertainment (a quarter), but it also tries to restore the balance with a serious attempt to offer different types of information programmes (one fifth of the global output).

VTM's success story strongly contrasts with the short, but difficult history of its second channel. Ka2 was launched as a strategic response to VT4, with a unique and sometimes amazing programme offer. For more than 6 months the second commercial channel scheduled a quite innovative mixture of the best of British and American fiction series and movies, larded with often strong news, current affairs programmes and talk shows. The audience however did not follow the project, and the channel's market shares (a couple of per cent) stayed far below the projected 7%. After a few months the advertisers withdrew support for the channel, which led the VTM shareholders to exert pressure on the central programming staff to change strategy. At the end of 1995 Ka2 died and a new channel was born: Kanaal 2 looked very familiar to the general audience, with a strong popular and entertainment-based programme offer. VTM's second channel closed the year 1995 with a market share of 3.9%. For the first time in its existence, the commercial station as a whole made big financial losses, leading to an open dispute between the shareholders[3] and the management - which finally ended with the withdrawal of the programming directors at the beginning of 1996.

The multi-channel environment has severely weakened the position of the BRTN, with the year 1995 as the public broadcaster's '*annus horribilis*': in 1995 the Flemish PSB lost the contract with the national football league, struggled with financial problems and its market shares seemed to skydive (see Table 3). After a critical McKinsey report, a new legal structure was created in December 1995 in order to strengthen the position and the autonomy of the BRTN. In this 'mini decree' it was stated that the public broadcaster had to become less bureaucratic and politicised, and to be led in a more professional way. One of the most visible changes at the beginning of 1996 was that the old (politically installed!) management was fired and replaced by a small, more independent staff of three people, headed by a delegate manager with hard private enterprise experience. The new decree also defined a vague mission statement, stressing the need for reaching the wide audience. After a short running-in period, the new powerful managerial staff quickly changed all leading positions, mostly engaging very experienced and successful broadcasting professionals.

Critics find it quite remarkable that most of these people were previously labelled as politically liberal-conservative, and that they all previously produced very popular entertainment programmes. Many fear that the programming style of both BRTN television channels will soon move in a more entertainment and

ince 1993 some 45% of TM's capital is controlled y the major Dutch ublishing and multimedia ompany VNU (Verenigde ederlandse Uitgevers). 5% is held by local print edia actors and financial roups popular direction. In recent years, TV1 and TV2 had followed a complementary programming strategy. The first channel tried to reach a wider audience, with lots of fiction (46% in 1995, mostly imported from the UK and US), news and information (22%), and entertainment (10%). The second channel was profiled as a fiction, sports and information channel, often directed to more specialised niche audiences. However, TV2 lost a good part of its audience share and became the smallest channel.

The channel whose future is most unpredictable is VT4. During the first months of its existence, VT4 had major difficulties in attracting its core audience (urban, well-educated men), so that it stayed well below its target of 10%. By the second half of 1995, the channel's strategy and image had been completely transformed, and now VT4 has an American style, with lots of series and program formats from the USA, and with a small proportion of local programming (less than 10%). As an entertainment-oriented channel, VT4 has been reducing its information output, and even decided in July 1996 to cut news. One of the 'hidden' secrets behind VT4's growing success has been the scheduling of late night erotic series. The result of such aggressive scheduling has been an increasing audience share (7.7%), but difficulties in attracting a really loyal audience.

The complementary French-language television scene

One of the outcomes of recent Flemish television history is that the local industry is tending to become more fragmented as a result of growing competition, although the local scene is still dominated by a couple of big stations. Another result is that local stations have been quite successful in reducing the appeal, audiences and advertising expenditures for foreign stations. Recent television history in the southern part of the country, however, shows a completely different picture.

Station/ Channel	Date of creation	Ownership and funding	Viewers	Profile
BRTN (two channels: TV1 and TV2)	1930 (as NIR), 1960 (as BRT, later BRTN)	state; funded by public means (79.1%), advertising (only on radio: 10.3%), sponsorship (2.1%), and other commercial activities (8.5%)	25.3% (21% for TV1, 4.3% for TV2)	Mixed programme offer, including information, fiction, entertainment. Sport on TV2
VTM (two channels: VTM and Kanaal 2)	1989; Ka2/ Kanaal 2 in 1995	Commercial, owned by the Dutch VNU (44.3%) and VMH (55.5%); totally funded by advertising and other commercial activities	48.2% (41.1% for the first channel, 7.1 for Kanaal 2)	Popular programmin with VTM as family channel: entertainme imported (US) fictio local soaps and sitcor shows, but also information
VT4	1995	Commercial, owned by SBS (via EBS) and VUM; totally funded by advertising and other commercial activities	7.7%	Strong emphasis on US action and adventure series and shows

Table 4: Principal Dutch-language stations

	1990	1995	1996 (March)
RTBF1	25.2	20.3	18.6
Télé 21	4.4	2.8	2.4
RTL-TVi	22.9	22.6	22
Club RTL	-	3.2	2.9
TF1	21.2	14.1	13.5
Fr2	10.6	8.8	9.5
Fr3	7.6	7.1	7.5
Canal +	1.1	0.9	2.5
Others	7	20.2	21.1

Table 5: Market shares of the main stations in the French community

First of all, it is clear that the Walloon television industry is subject to fragmentation and competition from foreign stations. Audience shares indicate a very divided market, with no channel attracting more than one quarter of the audience (Table 5). The French-language television scene can be characterised as a small, restricted and open market with a major part occupied by foreign stations. Some authors have even portrayed the Walloon broadcasting scene as a fragile, dependent and complementary market (e.g. Lentzen & Legros, 1995: 20, 41).

French television channels, led by TF1, still attract nearly a third of viewers. Scheduling on the leading local generalist channel, RTL-TVi (two-thirds owned by the major European transnational media company CLT, and financed entirely by advertising), is dominated by fiction (65%) and entertainment (13%), with the proportion of news and information (12%) falling recently. Most of its fiction output (70%) is American, while a quarter is of French origin. RTL-TVi's average market share tends to be quite stable around 22%.

At its February 1995 launch, Club RTL's evening schedule was built around a particular theme, with a popular movie as the major attraction. After a couple of quite successful weeks, the formula seemed to bore the public, and Club RTL did not reach its modest projected audience share of 5%. Its early evening offer, however, mainly children's and youth programming from American and Japanese production centres, seems to be more attractive. In June 1996, the RTL-TVi direction decided to increase the output of its second channel, and to schedule more local programming.

From a European viewpoint, the French-language RTBF could be considered as an interesting example of a traditional public broadcaster, having great difficulties in living up to its ambitions and its PSB-mission, due to fierce external competition, its problematic financial situation, and its heritage from the past. The picture of the RTBF as a heavily bureaucratised, politicised, overstaffed and ponderous institute is probably only slightly unfair. Since 1989, under new management and now taking advertising, a slimmer RTBF has seen better financial results, but a falling market share. Unlike its commercial rivals, the French-language public broadcaster, which depends mainly on public funds (6031 million Belgian Francs, or 67%), still tries to offer a mixed schedule, inspired by a traditional PSB-philosophy. Besides fiction (31%), entertainment (7%) and sports (11%), it devotes most of its time to news and information (34%). Critics fear that the growing dependency on advertising and the lack of political will to fund the public broadcaster will drive the RTBF programming strategy into the most predictable one of imitation.

Station/ Channel	Date of creation	Ownership and funding	Viewers	Profile
RTBF (two channels: RTBF1 and Télé 21)	1930 (as INR), 1960 (as RTB)	State; funded by public means (67%), advertising (19%), sponsorship (4%), and other commercial activities (10%)	21% (18.6% for RTBF1 and 2.4% for Télé 21)	Mixed programme offer, stressing information, fiction, discussion programmes. Sport reserved for Télé 21.
RTL-TVi (two channels: RTL-TVi and Club RTL)	1985; Club RTL in 1995	Commercial, owned by CLT (66%) and Audiopresse (34%); mainly funded by advertising (86%) and other commercial activities	24.9% (22 for RTL-TVi, 2.9% for Club RTL)	Focusing on fiction and shows; often imports from the US and France

Table 6: Principal French-language channels

Conclusions

The history of Belgian television shows how media reforms clearly reflect political and societal changes. Rooted in a traditional PSB-tradition within a strong unitary nation-state, the Belgian television scene slowly grew into a federalised system along cultural/linguistic lines. Since the 1960s both Belgian television industries developed quite different programming styles, working practices and ethos, although some common structural influences can be observed. Besides the threat to local monopolies from cable technology, there has been the dominant political force. The party political and ideological pressure eventually undermined the credibility of the public broadcasters as a forum for broad public discussion.

Along with the decreasing political control over broadcasting since the late 1970s, the public broadcasters have been ever more threatened by external competition (cable) and political-economic pressure to liberalise the television scene. Compared to other countries, Belgium, and especially the northern Flemish part, was fairly late in organising internal competition. After a first wave of commercial channels in the late 80s and a second one in the middle of the 90s, the viability and legitimacy of the local PSB-stations within the multi-channel scene is more than ever under attack. The most obvious solution has been to revitalise the public service by applying commercial solutions and criteria. The danger is that fulfilment of the role of a truly public service broadcaster will be constrained no longer by political pressure, but rather by viewing figures and economic profitability.

Principal organisations/abbreviations

BRTN	Belgische Radio en Televisie - Nederlandstalige Omroep: Dutch-language Belgian PSB of the Flemish community
CLT	Compagnie Luxembourgeoise de Télédiffusion
EBS	European Broadcasting System
F2/F3	France 2 and 3
INR/NIR	National Institute for Radiodiffusion
Ka2/Kanaal 2	second VTM-channel
RTBF	Radio Télévision Belge de la Communauté Française: French-language Belgian PSB
RTL-TVi	Radio Télévision Luxembourgeoise, Télévision Indépendante: French-language commercial television station
SBS	Scandinavian Broadcasting System
TF1	Télévision Française 1: French commercial station
VTM	Vlaamse Televisie Maatschappij: Flemish commercial television station
VT4	Viking Television 4: Dutch-language commercial station operating under a British licence
VUM	Vlaamse Uitgeversmaatschappij: most important newspaper publisher in Flanders

References

Biltereyst, D. (1992), 'Language and Culture as ultimate Barriers?', *European Journal of Communication*, 7(4), 517-540.

Biltereyst, D. (1995), *Hollywood in het Avondland. Over de afhankelijkheid en de impact van Amerikaanse televisie in Europa*, Brussels, VUBPress.

Biltereyst, D. & Burgelman, J.C. (1995), 'Belgium', in I. Hamm & F. Harmgarth (eds.) *Television Requires Responsibility. International Studies*, Gütersloh, Bertelsmann, pp. 69-102.

Burgelman, J.C. (1990), *Omroep en Politiek*, Brussels, BRT Uitgeverij.

Burgelman, J.C., Biltereyst, D. & Pauwels, C. (1994), *Audiovisuele Media in Vlaanderen. Analyse en Beleid*, Brussels, VUB Press.

CF/Communauté Française (1993), *Annuaire de l'audiovisuel*, Brussels, Ministère de la Communauté Française.

Jongen, F. & Stephany, P. (1990), *Les révolutions de 89*, Brussels, Crisp.

Kerki, M., Philippaerts, P. & Van Soest, J.C. (1995), *Openbare omroepen morgen*, Brussels, Masereelfonds.

Lentzen, E. & Legros, M. (1995), *Les télévisions en communauté française*, Brussels, CRISP.

Libert, M. (1992) VTM. *De Euforie voorbij*, Leuven, Kritak.

McQuail, D. & Siune, K. (1986), *New Media Politics*, London, Sage.

Media Marketing (1996), *De Odyssee van de televisie in Belgiâ*, Brussels, Dupuis.

OEA (1996) *Annuaire Statistique 96*, Strasbourg, OEA/Observatoire Européen de l'Audiovisuel.

Robillard, S. (1995), *Television in Europe: Regulatory Bodies*, London, J. Libbey.

Siebelink, R. Biltereyst, D. & Burgelman, J.C. (1994), 'Vlaanderen', in N. Van Zutphen & J. Nootens (Eds.) *Nederlandstalige en Afrikaanstalige Media*, Brussels, VUBPress & Flemish Parliament, pp. 221-277.

Witte, E. (1994), *De Breedte van het Scherm*, Brussels, Icarus.

Television in Greece
Anna Syngellakis

A brief profile of the Greek state and society
Greece, a small country at the southern end of the Balkans, has a population of 10.8 million people (1991 census) one third of which is concentrated in the Athens metropolitan area. Physically, Greece is characterised by a mountainous mainland and numerous islands. The population is remarkably homogeneous: the overwhelming majority speak the Greek language and belong to the Greek Orthodox Church.

Greece is marked above all else by a turbulent history spanning the centuries. The modern Greek state was established in 1830 following a war of independence against the Ottoman empire. A common reaction to many failings of the contemporary Greek state and society is to attribute them to the legacy of the four century long subjugation to the Ottomans which meant that Greece was cut off from the historical events that shaped modern Europe. The Greek people had to make a difficult transition: from being ruled by the Sultan, and having therefore continuously to devise ways of gaining some control over their lives, to adjusting to a centralised administration complete with an imported King imposed on the newly independent state by the super powers of the time. The pattern set then of a cumbersome, dysfunctional state has persisted. And, if one is allowed some degree of generalisation, one would say that Greeks typically have a highly individualistic character and feel a deep mistrust for the state while making every effort to benefit from state activities.

These characteristics of the Greek state and society had a significant degree of influence on the way Greek television broadcasting developed in the late 1960s. Moreover, the political realities in Greece in the post World War II era were particularly important in shaping Greek life at all levels. A key characteristic of Greek radio and television is that they were established during turbulent and undemocratic periods of Greek history.

Origins and development of Greek television
Television broadcasting was introduced on a limited, experimental basis in September 1960 (Dimitras, 1992: 90). The first steps to establish a regular television channel were taken in 1966. The early years of Greek television and the adoption of the first law to regulate its operations coincided with a dark period in Greek history, the 1967-1974 military dictatorship. These oppressive years were the culmination of a long period of turbulent politics. Greece fought heroically against the Italians and the Germans in World War II but was devastated after liberation by civil war followed by right wing governments supported openly by Great Britain and the USA. The oppressive political climate of ultra strong anti-left policies was naturally reflected in the radio and press.

The National Radio Foundation, EIR, was founded in 1945 but did not acquire a legal framework until 1953 (Law 2312/1953). This established EIR as a Legal Entity of Public Law with a monopoly over all electronic media. When television broadcasting was established in the late 1960s it came under the state broadcasting authority, which was changed into a state controlled joint stock company, re-named Greek Radio and Television (ERT).

The state monopoly over radio and television broadcasting was, however, subject to a unique exception. Since 1951 the Greek armed forces had had the legal right to install and operate radio or television stations in order to serve the needs of the armed forces regarding information, education, entertainment and, in times of war, in order to strengthen the spirit of the nation. In the post-civil war, right-wing political environment, the Greek armed forces indisputably played a key political role and built for themselves an extensive broadcasting service, YENED, which was independent of, and antagonistic to, EIR and then ERT. This unusual state of affairs continued till 1982 when YENED was transformed into a civilian, state-owned broadcasting service, ERT2. Finally in 1987 ERT1 and ERT2 merged into one company, ERT.

Up until the mid-1980s, the Greek media operated in a predictable, fixed setting: the press was in the hands of traditional publishers; radio and then television were controlled by the state. By 1987 the winds of change had brought fierce competition between the established media players and new players claiming a piece of the action.

From state monopoly to deregulation

Greek radio and television since the very beginning had been functioning on the basis of the principle that broadcasting frequencies are subject to the monopolistic control of the state. This principle was based on Article 15 of the 1975 Constitution which provided for the direct state control of radio and television which had the responsibility for 'the objective and fair broadcasting and news, as well as of products of literature and of the arts' and 'securing the programming quality required by their social mission and the cultural development of the country'.

Greek governments chose to interpret 'direct state control' to mean a state monopoly of the electronic media. They even stretched this interpretation to mean that radio and television should operate in a culture of extreme pro-government reporting. Social reaction to this biased diet of broadcasting brought mounting pressure for deregulation. A notable movement was led by a group of intellectuals called Channel 15 after the constitutional article.

A decisive turn towards deregulation occurred in 1987 when various mayors, loyal to the conservative opposition party, established radio stations. The socialist government was pressurised to pass legislation making it possible to establish municipal non-state owned radio stations. This development in turn opened the way for private and municipally owned television channels. A local television channel was soon established by the mayor of Thessaloniki. This direct action, representing an open challenge to the official interpretation of the law on broadcasting, forced the issue. At first the government sought to enforce the law by taking the mayor to court. The case eventually reached the European Court of Justice in Luxembourg which ruled that the Greek legislation granting to the state a monopoly over radio and television broadcasting was in breach of the Treaty of Rome. In 1989 the coalition government in power recognised the *de facto* situation and opened the field of radio and television broadcasting to interested parties, thus limiting ERT's monopoly. A National Radio and Television Council was established to regulate broadcasting activities. The first two private television channels, Mega Channel and Antenna TV, started broadcasting at the end of 1989, followed by numerous others. Amazingly, none possessed the licence required by law. This anomalous situation will be explained later. Suffice it to say here that in the context of Greek realities this irregularity was not a unique phenomenon.

As one can see, deregulation of the Greek broadcasting media was not the result of a well thought-out strategy with appropriate research, consultation and debates to establish options and policies. The Socialist government at the time responded in an *ad hoc*, reactive manner when both internal and external political pressures came to a head and forced the issue. It would be fair though to take into account the fact that Greece had, in a relatively short time, to change dramatically many aspects of public life and adjust to the norms of the European Community: after a long period of political turmoil, the return to democracy and the accession to the European Community happened in the space of seven years (1974-1981). There was therefore precious little opportunity for a smooth evolution.

The internal pressures towards deregulation were linked to public reaction to excessive state control. In fact the circumstances of the establishment and evolution of Greek broadcasting are a reflection of the peculiarities, tensions and ambiguities of the Greek political system. It has been argued (Mouzelis, 1979) that the strongly centralised nature of the Greek state is the result of the fact that, in contrast to other western countries, Greece acquired a parliamentary system of governance ahead of any industrialisation of its economy. The consequence was that the political and economic system had not developed self-regulating mechanisms and that civil society remained weak. The Greek state assumed a dominant, paternalistic role with extensive functions at the ideological, political and economic level and decisive control over all manifestations of public life. State intervention, censorship, overt or covert, and general control over the media was therefore a natural exercise of state power. In practical terms state power was wielded by the government of the day. A commonplace occurrence illustrating vividly this state of affairs was that any change in the government and in the balance of power meant a parallel change in the leadership of radio and television services. Even during any given government's term in office executives in the state media had to toe the government line on penalty of dismissal. Many appointments were short-lived. For instance in 1981-89 ERT had thirteen chairmen and directors, and sixteen news directors with an average term in office of about eight months.

The paternalistic links between the government and key media executives meant that reporting echoed official government policy on issues. An attempt by the Socialist government in 1985 to promote impartiality and pluralism of opinion in the state broadcasting media was abandoned after only two and a half months. Apparently they also found it useful to exercise tight control (Papathanassopoulos, 1990: 391).

Following the 1974 democratisation, this state of affairs was out of tune with public sentiment which was supportive of pluralism, openness and independence in public life in general and especially in radio and television. External pressures were brought to bear by the wider influence in the 1980s of the ideology of deregulation, the changes experienced by the media of most European states associated with rapid technological changes, and naturally the policies of the European Community. The EC's 1989 Directive, known as Television Without Frontiers, promoted neoliberal policies in order to open up the broadcasting market in the member states to competition. The incident with the mayor of Thessaloniki mentioned earlier illustrates the influence of Community law on a member state's broadcasting policy.

The deregulation of television broadcasting brought on to the scene a number of private interests mainly from publishing, especially press, or entrepreneurial backgrounds, not unlike the pattern elsewhere in Europe. The situation developed following the pattern familiar in press and radio: too many channels compete for limited sources of commercial income and their viability in most cases is dependent on state subsidies, soft bank loans or the personal wealth of their owners.

Impact of deregulation

Indisputably state television suffered a major set-back from deregulation. State intervention continued as before, audiences went down and income from advertising suffered accordingly. Radio broadcasting and the press were also hit by the advent of private television channels. A major anomaly that came about is that only the state television channels were operating legally while the two big private channels, Mega and Antenna were in a semi-legal situation because it was understood that they had the consent of the government to operate. In addition to these channels there was a chaotic situation with many illegal stations and even more pirate stations which did not even seek to obtain a licence. It has been estimated that more than 100 television channels and at least 1500 radio stations are in operation without being regulated in any way (Heretakis, 1993). What characterises the situation is the non-regulation of deregulation or, to put it differently, a deregulation carried to the extreme: a deregulated deregulation. This anarchic state of affairs seems to be a constant reality in the Greek broadcasting landscape. The impact of deregulation on the market share of television channels can be shown in Table 1(*To Vima*, 13-6-93).

The decline of the state television channels in Greece is quite dramatic but it is not out of step with what has been happening elsewhere, although it would appear that in Northern European countries this decline is less marked than in Southern Europe (*To Vima*, 13-6-93). In 1991 the Greek government commissioned the BBC to study the problems of Greek state television and make proposals for its future. The report found that state television was in effect bankrupt and proposed draconian measures of re-organisation which were met with stiff resistance from the ERT staff and the opposition party. Various plans have been put forward: that ET1 should focus on information and entertainment and ET2 on culture and education; that excessive staffing should be drastically reduced and employees re-deployed.

Undoubtedly deregulation was a watershed for Greek broadcasting. But it also offered opportunities to formulate a national broadcasting policy, opportunities which unfortunately have not been grasped.

	ET1	**ET2**	**Mega**	**ANT 1**	**Satellite stations**	**Video**	**Others**
1989 (before deregulation)	37.7	24.3			14.2	10.3	13.5
1989 (after deregulation)	27.1	24.3	35.3		10.7	5.8	8.5
1992	10.5	5.9	33.6	30.5	1.3	3.6	14.7

Table 1. Percentage market share

Key features of the legal framework regulating television broadcasting

The starting point for any discussion of the legal framework is to state that there is no lack of adequate legislation. The problem is the lack of implementation. At present the legal framework governing television broadcasting is in a phase of transition. New legislation, law 2328/95, was adopted in September 1995. It is widely accepted in Greece that the chaotic environment in which the media operated was in need of a new legislative regime. At the same time doubts are expressed as to whether the new law will be implemented in practice or whether it will remain 'on paper', like the previous ones. As ever, there is talk of the 'political cost' of displeasing influential media barons that no one wants to pay (Nathanail, 1995; Halvatzakis, 1996).

Identifiying the main features of the legislative framework

The establishment of the National Broadcasting Council (ESR)

According to law 1866/1989 ESR exercises direct state control over radio and television broadcasting in order to safeguard the freedom and plurality of expression, the observance of journalistic codes of practice and the objectivity and quality of radio and television programmes. ESR has its own secretariat and budget and its members are personalities drawn from the arts, sciences, politics or technology. Initially there were eleven members but in 1990 their number was increased to nineteen for political reasons. These political reasons, and the degree of ESR's independence, become apparent when one considers that the members of ESR are nominated by the three main political parties and by various professional associations and interest groups, such as the Association of Press Editors and the Greek Association of Engineers but also the Church, the Greek Court of Auditors and the Greek Chamber of Commerce. ESR members have been criticised as unable to distance themselves from the bodies that nominated them in order to establish in practice the independence of the ESR (Papathanassopoulos, 1992).

ESR is charged with the task of issuing codes of practice governing journalistic practice, programmes and advertising, of granting licences for cable television to certain residential areas and traditional villages, of giving to the responsible Minister its consent for the granting of licences to local radio stations, and of scrutinising the financial position of applicants and giving its favourable opinion to the Minister regarding the granting of licences to establish private television channels.

The ESR has often been criticised as an ineffective body, in practice dependent on the government of the day, whose members do not always have the technical background or understanding to grasp the rapid changes in broadcasting. As a result the ESR has not always been able to keep up with the explosion in radio and television broadcasting and has been notorious in delaying its recommendations to the government on applications for licensing. It is seen as at least partially responsible for the operation of unlicensed television channels, a commonplace phenomenon of Greek broadcasting.

Licensing of private television channels

According to the law adopted in September 1995 private television channels must be licensed by the government in compliance with the requirements of the law. Applications are invited every September and also whenever appropriate. Licences are valid for a period of 4 years, unless revoked by the government, and are renewable. The law stipulates that licences are granted in the public interest and their use amounts to a public function. It is stipulated further that licensees have a duty to provide objective reporting, plurality of opinion, and to promote cultural and artistic programmes. Sanctions for breach of these duties range from an initial warning to financial penalties or the revocation of the licence.

Media concentration

A company is allowed to hold only one licence for the establishment and operation of a television channel or to participate only in one other company holding such a licence. Any natural or legal person is allowed to be a shareholder in only one company holding a broadcasting licence and only up to 25% of the company's capital. The restriction is extended to the spouse or relatives who do not have financial and business independence from the person in question. The law is very detailed in this aspect and imposes a number of criteria to judge the person's eligibility. Furthermore one is not allowed to have interests in more than two separate forms of media (e.g., television and press or television and radio). These provisions of the Greek law are considered among the strictest in the European Union (*To Vima*, 13-8-95).

Advertising

A fundamental condition of the licence to operate a television channel is that the provisions regulating advertising are observed. The provisions apply Directive 89/552/EEC and they are quite detailed. The main points are: advertisements must be easily recognisable as such and must be clearly distinguishable from the main programme. Any indirect or subliminal advertising is prohibited. Television advertising must not offend against human dignity nor discriminate on grounds of sex, race, religion or nationality. It must not encourage behaviour detrimental to human health and safety and the protection of the environment. Minors are protected by extensive rules prohibiting any advertisements which, inter alia, expose them to moral or physical risks or exploit their inexperience and encourage them to buy goods or services. As a rule advertising may not cover more than 15% of daily broadcasting. Films may be interrupted once in every 45 minutes and advertising may last up to 9 minutes. Other programmes may be interrupted for up to 4 minutes.

Protection of citizens and consumers

The law provides a right of reply to persons whose honour, personality, personal or professional reputation are prejudiced by any television or radio programme. This right is also given to political parties and trade unions. Other rights under civil or criminal law are applicable. Consumer protection provisions are also applicable.

Protection of the Greek language

Radio and television are under a legal duty to ensure the correct use of the Greek language and to transmit programmes in the Greek language up to 25% of their total broadcasting time, in addition to news bulletins, sports reports and advertising. Another legal obligation, not observed in practice, is to broadcast once every six months a series of at least 15 programmes, each 30 minutes long, promoting the correct usage of the Greek language or its teaching to foreigners or to illiterate persons.

TV penetration

Television penetration has reached 99% in the last few years, while 38% of households have more than one set; 46% of households possess a video cassette recorder (*TV Year-book 1994-95*).

TV advertising

In 1995 the 8 larger channels (MEGA, Antenna, ET1, ET2, ET3, Sky, Star and New) broadcast over 3,484 hours of advertisements. Total advertising expenditure on mass media, including radio and press, reached 408,085,598,000 drachmas. Antenna absorbed 107,676,016,000 drachmas followed by MEGA with 85,120,570,000 drachmas (*To Vima*, 7-1-96).

Channel name	Date of creation	Ownership and funding	Viewers	Profile
ET1 & ET2	1966, merged 1987	State; state budget, fee incorporated in electricity bill and advertising	just under 10%	News; information; art and culture; children's programmes; films and series
MEGA	1989	private; owners include major newspapers; funded through advertising	combined share with ANTENNA 52%-56%	Light entertainment, news, information; equal share of Greek and foreign programmes
ANTENNA	1990	private; funded through advertising	combined share with MEGA 52%-56%	Light entertainment, news, information
STAR	1993	private; funded through advertising	18%	Emphasis on foreign and Greek series and films; news
SKY	end 1993	private; funded through advertising	8-10%	Emphasis on foreign and Greek series and films; news
Channel 5	1994	private; funded through advertising	NA	News and information. More foreign than Greek programmes
FILMNET	1992	subscription (7,500 drachmas per month)	110,000 subscribers	Mainly foreign films; sports
Seven X	1992	private	Not available	Foreign b/w art films, less emphasis on news
New Channel	1990	private	Not available	Entertainment: foreign and Greek films; news and information

Profiles and market shares of Greek television channels

Profiles and market shares of Greek television channels

Although the state television channels, ERT1 and 2, put substantial effort into ensuring quality programme making, their combined market share is small compared to that of the private sector, dominated by MEGA Channel and Antenna TV. Both provide a balance of Greek and foreign programmes addressed to all categories of audience. The focus of MEGA Channel is on films, series, light entertainment (talks shows, variety shows) and news. It broadcasts three extended information programmes and four daily news bulletins, compared to seven of each on Antenna TV. Star channel revamped its programmes at the start of 1995, emphasising foreign series and Greek and foreign films, and climbed to third position, while SKY channel lost ground. Seven X targets the middle and upper socio-economic groups. Channel 5, which has links with the French channels France 2 and France 3, broadcasts one extensive news programme and six three-minute long news bulletins as well as four information programmes daily: programmes are 43% Greek and 57% foreign. Filmnet, the only subscription channel, has secured exclusive broadcasting rights to many sports events and claims to have 110,000 subscribers. A new subscription channel, Super Sport, was established in August 1996 broadcasting mainly sports events while Filmnet concentrates on films. The new channel, under the name of K-TV, also provides children's programmes in the mornings. Finally there are a number of other channels which collectively share about 10% of the market and have more specific audiences, such as 902 Left which is linked to a specific party.

There is no cable television in Greece. Satellite dishes are not widespread but ERT transmits major foreign channels, such as CNN, MTV, RAI etc., which are thus widely received.

In conclusion, Greek television programming is dominated by information and news programmes followed by films and series. The private channels devote up to 75% of their total air time to such programmes while arts, culture, sports and children's programmes are less well catered for. In public television the programming mix is almost the. opposite.

Key legislative texts

Law 2312/1953: EIR established as a Legal Entity of Public Law with a monopoly over all electronic media.

1975 Constitution: Direct state control of radio and television (Article 15).

Law 1730 of 11-18 August 1987 re the establishment of the Greek Radio-Television Ltd. (ERT -AE).

Law 1866/1989 on the establishment of the National Broadcasting Council and the granting of licences for the establishment and operation of television channels.

Presidential Decree 572/4-8 December 1989 re the procedure in the National Broadcasting Council re the granting of licences for local television.

Presidential Decree 573/5-8 December 1989 re the appointment and responsibilities of the members of the National Broadcasting Council.

Ministerial Decision no.22255/2 of the 3-5 November 1990 re widening the membership of the National Broadcasting Council.

Presidential Decree 236 of 10/16 July 1992 re the application of Directive 89/552/EEC.

Law 2328/1995, September 1995, re the regulation of media of mass communication.

Abbreviations

EIR	National Radio Foundation
ERT	Greek Radio and Television
ET	Greek Television
ESR	National Broadcasting Council
YENED	Broadcasting Service of the Greek Armed Forces

Key dates in Greek television

1945	National Radio Foundation (EIR)
1951	Greek Armed Forces obtain legal right to operate radio or television stations, the YENED.
1960 (Sept.)	Experimental TV station at the Annual Trade Fair, Thessaloniki.
1966	First steps to establish a regular television channel.
1967-74	Military dictatorship
1981	Greece accedes to the European Economic Community
1982	YENED, the armed forces station, becomes a civilian channel, ERT2.
1987	ERT1 and ERT2 merge into one company, ERT.
1987	First steps towards deregulation.
	ECJ judgement
1989	ERT monopoly ends. National Radio and Television Council established (ESR)
	Mega Channel and Antenna TV established.
1990	ESR membership widened.
1994	Filmnet established; only subscription channel.

Bibliography

- Legislative texts in Greek.
- Press reports in Greek, especially reports and surveys published in *To Vima*, an authoritative Sunday paper.
- Personal communications with TV channels and independent journalists in Greece.

Dimitras, P. (1992), 'Greece', in *The Media in Western Europe, The Euromedia Handbook*, Sage.

Halvatzakis, K. (1996), 'The law on the media in suspension' (in Greek), *To Vima*, 9-6-96.

Heretakis, M. (1993), 'Developments in the Greek Media' (in Greek), *The Manager*, no.38, February 1993.

Katsoudas, D. (1987), 'The media, the state and broadcasting' in Featherstone, K., and Katsoudas, D., *Political Change in Greece: Before and After the Colonels*, Croom Helm, London, 1987.

Mouzelis, N. (1978), *Modern Greece: Facets of underdevelopment*, Macmillan.

Nathanail, P. (1995), 'Will the new law put order to chaos?' (in Greek), *To Vima*, 15-10-95.

Papathanassopoulos, S. (1990), 'Broadcasting, politics and the state in socialist Greece', *Media, Culture and Society*, vol.12, pp.387-397.

TV Year-book, 1994-95 (1996), AGB Hellas.

Television in the Russian Federation

Paul Flenley

The history of television broadcasting in Russia is one of the transition from a centralised state-controlled system serving the ideological and political needs of the Soviet Communist Party to a diverse network of channels partly state and partly privately owned which are having to respond to a range of pressures both political and commercial.

The Soviet period

Soviet television had its origins in the 1930s with experimental transmissions in 1931 and the broadcasting of a regular service to 100 television receivers in the Moscow area in 1939 (McNair, 1991:48). After the war the expansion of television was seen as something of a priority. In the 1950s television centres were established in the republican capitals and other major cities and by 1960 there were some 4.8 million domestic television sets. By 1975 this had increased to 55 million (Mickiewicz, 1988a: 18). Colour transmission began in 1967. In 1994 over the whole of the former USSR there were some 93 million television sets reaching a population of over 291 million (Fedorovski, 1994: 187). This growth of access was paralleled by the growing significance of television in people's lives. Whereas in 1979 79% of the population were reported to rely on newspapers for their primary source of information, by 1992 television had become the main source (Tolz, 1993: 205).

In developing the network the Soviet state expected television to serve the same functions as other media. Ultimately it was geared towards mobilising the population behind the policies of the government and Party. In view of the fact that the same images and messages could now be broadcast simultaneously to peoples in all parts of the Soviet Union from Estonia to Uzbekistan, television was an important instrument for binding Soviet society together. It was a powerful mechanism for the integration of a diverse society of over 120 different nationalities.

The state's desire to manipulate television for its own ends was complicated by changes in the nature of the audience and by the nature of the medium itself. The 1960s and 1970s saw the development of a more educated, urbanised population with more leisure time. The Party's message had to respond to a more demanding public less easily satisfied by bland political ideology and boring presentation. In particular, people were looking to television to entertain them as much as to inform. Even in the later Brezhnev period, then, one can detect greater attention to presentation in programming and the screening of more popular entertainment features.

Television was also a much more complicated medium for the state to control than radio or the press in that it was more difficult to manipulate the audience's reaction. Soviet citizens could be told one thing in news broadcasts but there was little to prevent them making alternative interpretations about the actual images they saw on the screen. Even the most politically acceptable Western films or critical programmes about the West could not help revealing to the public visual images of the level of Western lifestyles and of the greater liveliness of American popular culture. Citizens were also able to see their political leaders exposed on screen: unlike newspaper images, television images could not be retouched. It was perhaps unfortunate for the political system that this opportunity to view leaders on a systematic basis coincided with the very period when that leadership was descending into rule by a gerontocracy. The sight of an increasingly incapacitated Brezhnev struggling to make speeches or present medals was undoubtedly a factor in promoting the cynicism which pervaded the years before Gorbachev. Equally, at least in the first years after Gorbachev came to power in 1985, it was a relief to most to be able to view a new leader, who was energetic and fit. Television could portray him moving around the country talking to people. For a time at least television portrayed the view that there was life in the old system and *Perestroika* (reconstruction) could succeed.

The policy of *glasnost'* (openness) launched by Gorbachev in the spring of 1986 was based on the realisation that the old methods by which the Party had communicated with the population were counterproductive and if anything were contributing to the crisis of legitimacy from which the system had begun to suffer. Television had to respond to a more sophisticated audience if the Party's message was to be listened to at all. Under the new openness many more themes were discussed than ever before (Yasmann, 1988). By 1989, such topics as strikes, nationalist unrest and the role of the KGB could be examined. Western politicians were interviewed in an uncensored form for the first time. 'Bridge' programmes were broadcast involving live discussions between audiences in the West and the USSR. Investigative journalism became very popular and such programmes as *Vzglyad* and *600 Seconds* attracted huge audiences. In addition, Soviet citizens could see officials being interviewed and the proceedings of the new Congress of People's Deputies and Supreme Soviet were shown on television. The sight of politicians arguing and even high officials such as Gorbachev himself being openly challenged had a dramatic effect in changing people's political consciousness. Television played a major role in destroying the fear which had underpinned Soviet society for so long. The message that came over was that it was now acceptable to engage in open criticism of the government and its policies.

Although this was clearly a revolution in the content of television broadcasting, *glasnost'* did not see the transition to totally free media. The fundamental role of television and the media in Soviet society still remained the same, i.e. to mobilise support for government and Party policy. *Glasnost'* was seen as a way of inducing support for the policies of *Perestroika*. In the first year after coming to power Gorbachev had hoped to make the system work through greater efficiency and in particular by removing obstructive and corrupt officials. Gradually he realised that it was the nature of the bureaucratic system itself which was the problem and in this context the media had a very important role to play in seeking to expose abuses by officials and putting pressure on the system from below. Given this function there were limits to *glasnost'*. Criticism of the very highest officials was discouraged and there were still restrictions on the reporting of certain issues. There was only limited reporting of the Nagorno-Karabakh conflict in 1988. Perhaps most dramatically of all, the failure of the media to respond to the Chernobyl disaster in 1986 showed the early inadequacy of the *glasnost'* policy and as a consequence provoked pressure for more openness from 1987 onwards. News of the disaster was only released two days after the event when foreign reporting had made a cover-up impossible, and it was not until a fortnight later that Gorbachev went on national television to address the nation on the subject (Benn, 1992a: 186).

Throughout the whole of the Gorbachev period the media was vulnerable to the whims of the state. The new approach of *glasnost'* was not given any legal acknowledgement until August 1990 when the Media Law abolished censorship. This applied except in cases of the disclosure of state or other legally protected secrets. It was up to the state to decide what constituted a secret. The Law also talked rather ambiguously of the need to prevent abuses of freedom of speech by the media. There was still therefore little in practice to prevent the government seeking to censor the media. Nevertheless, Glavlit, the chief censorship agency, was abolished on 1 August 1990 (Benn, 1992a: 178). Glavlit (the Main Directorate on Matters of Literature and Publishing) had been established in 1922 and had the task of prior vetting of all press articles and broadcasts. Its abolition, however, did not mean an end to state control of television content. Throughout the Soviet period overt censorship of the media was really the last line of defence in a complex system of state and Party control. Overall supervision of the media was exercised by the Party's Central Committee Secretariat through a series of departments. Chief among these was the Ideology department which had a Mass Media Sub-department. More detailed control of television was exercised via GOSTELERADIO (State Committee of Television and Radio) and via state control of news through TASS, the state news agency. As in every institution in Soviet society appointment of managerial personnel, editors, senior journalists was a matter of consideration for the Communist Party. People in senior positions would be Communist Party members and they would be subject to the discipline of the party cell in their organisation. The heads of TASS and GOSTELERADIO were themselves appointed on the approval of the Party's Central Committee Secretariat.

Even in the last year of the Soviet Union's existence, Gorbachev could use these mechanisms to try to exert greater control over the media. In 1991 the country seemed to be lurching into crisis. Conservative opinion was being more forcefully expressed within the Party and state and it was felt even by Gorbachev himself that *glasnost'* was getting out of hand. Many blamed the media for undermining the stability of society with its emphasis on crime and its keenness to dwell on the mistakes of the past. In this context Gorbachev set up a new body, the All-Union State Television and Radio Broadcasting Company. It sought to impose stricter editorial control over television and also bring the republican media under tighter central control (*Moskovskie novosti* No.7, 1991). Moscow television's coverage of the shooting of demonstrators by Soviet troops in Vilnius on 13 January 1991 showed the extent to which new censorship was being imposed. *Vremya*, the main news programme, reported that Soviet troops did not fire first, an assertion contradicted by foreign journalists who had been present (Tolz, 1993: 198).

The one major structural change in the organisation of television in 1991, the last year of the USSR, was the establishment of the Russian Television Channel (Rossiya). This was a response to the rising demand of Russian nationalists of both left and right that the Russian Republic within the USSR should have its own television channel in the same way that other union republics had theirs. In the political conflicts of 1991, the channel was to introduce an element of political pluralism in state television. While Channel 1, Ostankino, continued to present a bias towards Gorbachev as President of the USSR, Rossiya on Channel 2 championed Boris Yeltsin who had been directly elected President of the Russian Republic.

Post-Soviet television

The collapse of the Soviet Union in 1991 did not bring an end to state control of television. It is true that the Communist Party, and the local cells through which it had exercised its control over Soviet society, including television, were banned in the Russian Federation. However, local politicians and the heads of regional administration were still able to exert pressure on newspapers and television (Wishnevsky, 1994: 2). At the centre of politics, television became an important factor in the conflict which developed between Yeltsin as Russian President and the Russian Parliament in 1992-3. The conflict centred on a wide and complex range of issues including whether parliament or president should have ultimate authority and also the pace and consequences of market reform. People who identified themselves as democrats were aligned with both sides. However, Yeltsin was able to use his control of state television to portray the deputies in Parliament as an amalgam of nationalists and old-style communists who wished to return to the past.

In late 1992 Yeltsin established the Federal Information Centre, headed by an associate, Mikhail Poltoranin. Although on 27 May 1993 the Constitutional Court declared the Centre to be unconstitutional, it continued to function and interfered in the organisation of the state-run media. Poltoranin installed three pro-Yeltsin managers at the main state channel Ostankino. By now control over state television was becoming an issue in the political conflict. Parliament fought back with the establishment of the Federal Oversight Council which sought to bring appointments to the leadership of state television under parliamentary control and tried to interfere in programming. Needless to say, this was rejected by Yeltsin and also state television itself (Tolz and Wishnevsky, 1993).

During these months of crisis, television was the most pro-Yeltsin of all the media. After the President's formal dissolution of Parliament on 21 September 1991, a number of programmes were taken off the air, including *Parliamentary Hour*, Parliament's daily programme, and *600 Seconds*, the investigative programme produced by the right-wing journalist Aleksandr Nevzorov (Wishnevsky, 1993: 1). Yeltsin's political opponents were denied access to television and the state channels regularly presented news and programmes favourable to the president's views (Tolz and Wishnevsky, 1993: 14). Not surprisingly, the violent events surrounding the attack on the Russian Parliament in October 1993 included clashes between supporters of Parliament and government troops outside the Ostankino television centre in Moscow.

In the run-up to the December 1993 elections Ostankino, the state television Channel 1, appeared openly to favour the pro-government Russia's Choice bloc (Tolz and Wishnevsky, 1993: 15). As a result, state television became too closely identified with the unpopularity of the government and its market reforms, thus denting the impact of any television coverage and also boosting the surprise success of Zhirinovsky and the Communists in the elections. Realising this mistake after the election results, Yeltsin removed Bragin from the chairmanship of Ostankino and replaced him with Yakovlev, who as mastermind of the liberalisation process under Gorbachev was more acceptable to a broader range of democratic opinion. Other swift changes included the state channel Rossiya's being compelled to lease five hours of time to the commercial company NTV. During the whole crisis, independent television had provided more impartial coverage. In addition, Yeltsin issued a decree guaranteeing free access to information of a general nature and establishing a court of seven jurists to adjudicate on controversial issues of access to the media (Wishnevsky 1994).

However, in practice, the aftermath of the December 1993 elections did not see a complete change of attitude by the state. On 22 December 1993 Yeltsin replaced the disliked Federal Information Centre and the Ministry of Press and Information with two new government organs, the Federal Television and Radio Broadcasting Agency and the Committee on the Press, which are seen as having similar functions of interference. In the course of the war in Chechnya it has been difficult for the government to avoid trying to continue to exert pressure on the media. The commercial channel NTV sent two teams of journalists to the area three weeks before any other television station and made reports which contradicted those of state television and the government's own statements. As a result it was reminded of its reliance on government controlled facilities for the transmission of its signal (Mickiewicz, 1995: 165). In April 1995 NTV's popular satirical puppet show *Kukli*, the equivalent of British television's *Spitting Image*, became the target of criminal investigations by the federal prosecutor for its disrespectful portrayal of government figures.

In the run-up to the December 1995 elections, there was some attempt by the Central Electoral Commission to ensure fair treatment for the different parties. Nevertheless, according to Sergei Markov of the Carnegie Endowment for Peace, there was still political pressure on the state channels to favour pro-government figures. Once again, however, as in 1993, negative treatment of the Communists and Nationalists on television seemed to work in their favour in the polls. These elections exposed even more clearly the complex relations between Russian television and the voting public. Certain parties and figures that were relatively new to Russian politics such as Alexander Lebed did benefit from television exposure. However, the Russian Communist Party, still the most well-known and most established party, found it could rely better on newspapers. The fact that it drew on a more negative, anti-government vote meant that exposure of government figures on television if anything reinforced its message among its supporters (Nivat, 1996a: 35).

Television coverage of the June/July 1996 Presidential election was unashamedly biased in favour of Yeltsin both on state and private channels. The chairman of NTV, Igor Malashenko, was a member of Yeltsin's campaign team. Such bias cannot be wholly explained in terms of state pressure. Businessmen, now heavily involved in the financing of television, feared that a Zyuganov (Communist) victory would damage the economic reform process. More specifically in television they feared the possible removal of advertising revenue. Journalists generally foresaw greater restriction of freedom of expression if Zyuganov won (*Guardian*, 18 May 1996: 17). To some extent there were grounds for suspecting this. The new Parliament with its Communist/Nationalist majority had already tried to pass a law 'On Television and Radio Broadcasting' which would have extended Parliamentary control over the content of programmes on state television (*Moskovskie novosti*, No.12, 1996: 39). Where the stakes are seen as so high in terms not only of the development of the country but specifically in terms of the future direction of the media, it is difficult for television to remain wholly neutral in politics.

The commercialisation of television

The major development of the post-Soviet era has been the increasing influence of business on the whole of television broadcasting, including the state channels. 1993 saw the establishment of the first private national television broadcasting stations. In January, Edward Sagalev, formerly head of youth programming at Ostankino, started up TV-6 (*Nezavisimaya gazeta*, 10 Oct.1992). Later in 1993 NTV, the most popular private channel, was set up and led by Igor Malashenko, a former deputy at Ostankino (*Argumenty i fakty*, no.12, 1995). The large banking group MOST has the controlling interest in the company which owns it. In June 1996 Gasprom, the huge oil and gas combine, took a 30% share in the same company (*Moscow News*, No.23, 1996: 7).

Greater commercialisation has been a consequence of two main pressures. Firstly, by the beginning of 1994, the growth of the private sector in the economy as a whole had been such that business now felt itself in a strong enough position to begin to exert its influence more openly on a political level. One of the main avenues for this is television. Business feels the need to influence public opinion in its favour. This does not necessarily mean backing one particular party or personality but trying to create a political climate which is favourable to business interests (Blagovolin, 1995).

A second reason for business to step in has been the parlous state of government finances (Bogomolov, 1995b). Although the government realised the importance of television in forming public opinion it could not afford to finance even the two state channels to the level required. Ostankino and Rossiya were receiving only 30% of what they required in 1994 (Tolz and Wishnevsky, 1993: 12). In the first half of 1994 the Ministry of Finance halved the budget of Rossiya. By 1995 state television was in real crisis. Much of this crisis was caused by individual state institutions themselves having to work on a commercial basis and using their monopoly position to increase charges to other state institutions such as state television. In January 1994 the Ministry of Communications threatened to cut off all radio and television transmitters if radio and television stations did not pay the 67 billion roubles they owed (Wishnevsky, 1994: 3).

The immediate response of state television was to cut back on programming. Rossiya reduced broadcasting by two hours per day (Wishnevsky, 1994: 3). Channel 1, Ostankino, began closing foreign news bureaux and increasingly relied on old footage to cover foreign stories (Mickiewicz, 1995: 167). A more long-term solution for state television might have been an annual licence fee. However, in the absence of such an arrangement, state broadcasting had to turn to investment from the private sector.

In November 1994 the state Channel 1, Ostankino, was turned into a semi-commercial, semi-state concern..ORT (Russian Public Television) was established with 51% ownership by the state and 49% by private businesses. Chief among the latter were eight major banks and companies (*Argumenty i fakty*, No.12 March 1995). Many have viewed this merger of state power and private capital with apprehension. Oleg Poptsov, formerly chief of the Rossiya channel, warned of the unfair advantages ORT would have in being able to draw on the public purse as well as private funds. It would have the financial power to poach programmes and talent from its competitors (Bogomolov, 1995b). In addition, critics saw a new danger to media freedom posed by the control of large financial groups backed up by the power of the state.

For the investors involved, the attraction was not so much the chance to make profits through investment in television. In fact Boris Berezovsky, head of Logovaz, one of the main investors in ORT, complained that in its first year ORT made a loss of some 100 million dollars. The main attraction was the need for business as a whole to influence the political atmosphere. Channel 1 is the most popular channel, reaching 95% of the population and having a rating of 74%, ahead of the all other channels. Logovaz's involvement in Channel 1 was based on the belief that the country's main channel should be committed to support the general course of reform (Bogomolov, 1995b).

A whole number of private television production companies have also emerged which make and sell programmes to both the state and private broadcasters. Among the most successful are ATV, REN TV, 'ASS-TV', and Klass, all of which make programmes for ORT, and *Svezhii vecher* (Fresh Evening) which makes the programme *Good Morning* for Channel 5. NTV's *Kukli* programme mentioned earlier is actually made by the Dixi company.

As well as receiving the major national television channels, over 36% of the population are reached by a network of regional television companies such as Afontovo in Krasnoyarsk. Such stations can be private or state-run. Many of the local private stations co-operate to achieve economies of scale. One such venture is NVS (The Independent Broadcasting System) which is geared to the joint purchase and production of programmes for use by some 50 non-state television stations (Vul'chek and Muratov, 1995; Bogomolov, 1995b). Another feature of independent local broadcasting is the existence of highly localised cable television stations. These often serve just a small group of apartment blocks or a local area within a city. There are some 200 such stations in Moscow alone. They are usually financed by subscription and local advertising. One way of filling up the air time on such stations has been to show pirated foreign films.

The next most dramatic development in Russian television will be satellite channels. On 11 June 1996 NTV announced its intention to establish five satellite channels in Autumn 1996. The channels will be dedicated to sport, family, Russian films, foreign films and information. The ultimate aim is for ten channels by 1997 including ones dedicated to cartoons, variety, theatre and music. The MOST banking group has already bought two satellites and Gasprom joined in the ownership of NTV in 1996 in order to help finance the project. They hope to recoup the initial $170-180 million outlay largely through subscriptions in the first two years.

The state-run RTR company (Rossiya) has also announced that it intends to launch a sports channel, Meteor Sport. The main spur to these developments is the lack of space on existing terrestrial channels. NTV finds its eight hours of broadcasting on Channel 4 too little to permit the showing of its large stock of western films. Live nation-wide television coverage of sports events has been impossible so far on the existing network (*Moscow News*, No.23, 1996: 7).

Advertising

A major source of extra finance for both state and commercial channels has been advertising. Advertising on television existed even in the pre-Gorbachev period. However, it was very limited and rather crude in form. The aim was basically to let people know about goods available in state shops. As such, they were largely superfluous. Given the general shortage of goods, anything of decent quality was quickly bought up, syphoned off by those working in the relevant industry or allocated to the shops for the Party elite. The suspicion was that the real purpose of advertising was to shift goods which could not be sold, especially at the end of the plan period. The retail trade had their targets to fulfil like everyone else.

Under Gorbachev advertising increased and foreign capitalist companies were allowed to advertise on state television. Multinationals such as Pepsi took up the opportunity. This usually preceded the availability of the actual goods concerned (McNair, 1991: 87). The aim was to establish the familiarity of their brand names in a market which was bound to open up in the near future.

After the collapse of the Soviet Union, the opportunities for television advertising expanded enormously as commercial companies were set up both locally and on a federal level, especially from the beginning of 1993. The state channels also expanded their use of advertising to supplement their budgets and by 1994 they were taking the lion's share of advertising revenue - 38% to Ostankino (Channel 1) and 30% to Rossiya (Channel 2). The two largest commercial companies TV-6 and NTV took 3% and 12% respectively (*Moskovskie novosti*, no.31, 1995). For business the logic of advertising on the state channels was that Ostankino and Rossiya reached the largest audiences.

So far there has been little idea of targeting audiences in television advertising. One of the main consequences of this blanket coverage is that the Russian public is bombarded with images of goods and services which the majority of the population know they could never hope to afford. In the West, mass advertising is largely geared to influencing people's real choices or aimed at specific markets. For those millions of Russians, especially in the state sector, who have seen their real incomes drop and who face a future of employment uncertainty, the daily invitation to use financial services or buy expensive cars is little more than a cruel taunt (*Moskovskie novosti*, no.22, 1995). It reinforced the view that Yeltsin's market reforms had made a small number of people very well off and able to enjoy the goods displayed on television at the expense of the rest of the population.

For advertisers themselves there is the increasing realisation that saturation advertising does not necessarily pay off in terms of sufficient increase in sales to cover the costs of the mass advertising itself (*Moskovskie novosti*, no.31, 1995). The television companies have also recognised that there is a limited pool of advertising revenue and that in any case it does not provide a full solution to the overall problem of funding. Direct commercial investment appears the only answer. When Ostankino went semi-commercial and brought in private capital under ORT, it decided to impose a moratorium on advertising on Channel 1 in 1995 (Blagovolin, 1995). One other reason for the more sceptical approach to advertising has been the lack of any effective control over the validity of claims made by those companies advertising. The most dramatic example of this was the scandal over the investment firm MMM which was heavily promoted on state television in 1994 and then subsequently crashed with huge losses incurred by those persuaded to invest in the belief that the firm had the approval of state television.

The impact of recent changes

Increased commercialisation has meant that, as with all other areas of the Russian economy, television is even more open to corruption. This can take a variety of forms. Television journalists and producers on low salaries are vulnerable to bribes to push favourable reports or ignore sensitive issues. As Ostankino (Channel 1) moved to semi-private status, the fear was expressed that the nation's main television channel would now be corrupted. The murder of the television journalist and director-general of Channel 1 Vladislav Listyev in early 1995 was the most dramatic indication of this possibility. The suspicion was that his murder was connected with the control of advertising on Channel 1 and the subsequent decision to end it (*Time*, 4 March 1995). Even if the central authorities of a television company themselves are clean and attempt to stamp out corruption, they are hampered by the fact that much of a channel's output is often bought from independent production companies. In the case of the latter all kinds of deals over unofficial 'sponsorship', favourable reporting and placing of adverts can be done, which are impossible for the purchasing company to monitor (*Argumenty i fakty*, No.4, January 1993: 6).

One of the major consequences of operating in a commercial environment has been for television to try and pay more attention to the demands of its audience. In the Soviet period the priority was to fulfil targets and satisfy the Party and government. Little attention was paid to whether people actually watched the programmes, still less to whether alternative programming might have been more popular. The concept of ratings was unheard of. In the post-Soviet era the need to attract advertising and the involvement of business in ownership have made most channels more sensitive to programming and the need to maximise their audience. In this sense one can talk of the massification of Russian television for the first time. Whereas in the past the output of broadcasting was defined by the views of a small elite, now the orientation is towards the tastes of a mass audience. The consequences of this have not been entirely positive. At first with *glasnost'* there was no contradiction between massification and quality. People's thirst for new revelations, whether about the past or present, ensured that quality investigative programmes, documentaries, etc. could attract large audiences. To be a good journalist or television presenter was to have real social prestige in contrast to the cynicism accorded to journalism in the pre-*glasnost'* era. In the past few years, however, viewer preferences have veered away from programmes involving investigative journalism or politics to mass entertainment. Soap operas such as *Santa Barbara*, *Edera* and *Wild Rose* attract very high ratings of between 40% and 50% of the audience (Bogomolov, 1995a). The Moscow television company MTK began showing *Dynasty* on Channel 3 in Autumn 1995. One of the other most popular ways of boosting ratings is the game-show. *Field of Miracles* (*Pole Chudec*) is an enormous pull for Channel 1. The other channels have followed suit with their own versions such as *L-Club* on Channel 2 and *Blef-Klub* on Channel 5 (Bochkarev, 1995).

The advantage of such programmes for a strapped-for-cash industry is that they are cheap to make and attract the largest audiences. Many of these programmes are also bought cheaply from abroad, such as Mexican soap operas. Even where they are home-grown, the model is very often one taken from a foreign programme. One of the most successful imported formats is the talk show, literally '*tok-shou*' in Russian. Again Channel 1 led the way on this. Another way of attracting an audience cheaply is to be provocative. TV-6 transmits a number of highly popular provocative news programmes such as *Sharks of the Pen*, or *Scandals of the Week* (Bochkarev, 1995).

Attracting audiences has also meant the television channels having to change their image and the way they project themselves. The last few years have seen a move away from the rather staid presentation of the Soviet era. Television companies use computer graphics and music to push eye-catching logos, and television presenters are less formal than in the past.

Much of the content and the form of this revolution in television programming has been taken from the West, especially the U.S.A. To some extent this is a transitional phenomenon. The collapse of the Soviet Union and the move to a market economy meant that television had to adjust almost overnight to a radically changed environment. In particular, it had to respond to the demands of a mass audience for the first time, but in a period when finance was short. The American television broadcasting system had the most experience of serving a mass audience and not surprisingly it was to the models provided by it that Russia turned. In addition, domestic channels simply did not have the resources to respond to the needs of mass programming through home production. Cheap foreign products were a useful way of filling the airways and meeting demand at the same time (Vul'chek and Muratov, 1995). In addition, it was American companies who were the most forward in making the much needed foreign investment in Russian television both at a national and regional level (*Moskovskii komsomolets*, 20 July, 1995). To many nationalists this poses a threat to home-grown Russian talent and is seen as swamping the new Russia with Western culture and alien images. There have been calls for the state to intervene and at least impose some form of regulation to protect programming which is geared to the promotion of native culture (Rasumovskii, 1995; *Moskovskie novosti*, No.10, 1996: 38).

Part of the problem is the lack of effective regulation of the output of television. While journalists have complained of state intereference in political coverage there has as yet been no definition of the missions of the various channels. The main state Channel 1 run by ORT has the vague duty to serve the nation as a whole, but what this means in practice for the type of programming is not known. As Vladimir Sukhoi, an ORT journalist, has admitted, 'No one at ORT has a clear idea of what ORT should be.' (Nivat, 1996b: 41). Rather than seeking to establish viewer loyalty, programming policy seems to reflect the priorities of the particular personal or business interests in ORT. In addition, there is no regulation of the degree and nature of advertising even on the state channels. Moreover, as business moves into the media more and more there are no checks on the degree of ownership. Large concerns such as MOST and Gasprom are already significant players in the field.

Conclusion

In the period 1991 to 1996 Russian television has, like the rest of Russian society, undergone a revolution. It is now incomparably more diverse both in terms of ownership and content than it was in the Soviet period. In achieving this, television and the media generally have perhaps been the greatest success story of the Russian transition process. Not surprisingly, however, television still faces major problems, and not just of a financial nature. Given the political importance of the media in any modern society, Russian broadcasting has been unable to avoid playing a central role in the wider political conflicts going on in Russia. At one and the same time it has been key player and victim. In this context it has proved difficult for the concept of independent but public broadcasting to evolve fully. Television has come a long way from the state controls of the Soviet era. The quality of news reporting and the degree of scrutiny of those in power has improved immensely. Nevertheless, the government, businessmen and politicians still feel the need to bring direct pressures to bear on the media, especially while the fate of Russia still remains in the balance. In many ways this is part of a wider problem. As Russia emerges from the legacy of Stalinism it will take some time for the institutions and values of a stable civil society to be created. A television broadcasting system serving the public interest as distinct from those of the state or of vested interest groups must be part of that project.

Channel name	Creation date, privatisation	Ownership and funding	Viewers	Profile
Channel 1, ORT	Formerly Ostankino, founded 1931. Semi-privatised Nov. 1994	Formerly state owned. Now 51% state, 49% private, including 12 banks and companies, notably LOGOVAZ . Takes advertising.	74% rating, reaches 95% of population including many former republics.	News, serials, films, sport, talk-shows, current affairs.
Rossiya (Russian Television) uses Channel 2.	Spring 1991	State-owned by VGTRK company or RTR, also takes advertising.	57% rating, reaches 93% of population.	Similar output profile to Channel 1. Also runs "Russian Universities" on Channel 4
2x2, uses Channel 3	1993	Private, owned by TOO company which includes the MOST banking group. Takes advertising.	Largely Central Russia, Moscow area	Serials, films, advertising programmes.
Moskva or Moskovskaya programma, on Channel 3	1993	Private, owned by joint-stock company, MTK. Takes advertising	Largely Central Russia, Moscow Serials, Moscow	news, finance, business, music From 6.00 p.m.
NTV, uses Channel 4	Began in 1993 on Channel 5, then on Channel 4 from January 1994	Private, owned by TOO company. MOST has controlling interest. Gasprom 30% stake since June 1996.	68% rating, reaches 59% of population	Most popular private channel. Similar profile to Channel 1. From 6.00 p.m.
Petersburg" on Channel 5		State-owned	Received in central areas.	Serials, sport, children's TV, news. From midday
TV-6	January 1993	Private, run by joint-stock company MNVK (Moscow Independent Broadcasting Corporation). Takes advertising	42% rating, reaches 26% of population.	News, current affairs, serials, films, theatre reviews, finance.

Table of Principal Television Channels

Notes to table:

1. The lack of clear missions for television channels means that the overall profile for stations is often the same with some variations e.g. more local Moscow news on Moskva or more specific advertising magazine programmes on 2x2. However, there are differences in style between individual programmes on different channels e.g. opinion is that the current affairs programme *Itogi* on NTV is more incisive than similar programmes on Channel 1.
2. Given the size of Russia, the whole country does not receive the same diet of programmes. Combinations vary widely. Much of the country receives Channel 1, Russian Television, NTV and TV-6 and may also receive the 2x2, Moskva and St. Petersburg programmes. The regions receive output from other regional companies in addition to the above.

Key Dates

1917 (9 November)	First Soviet law authorising censorship, 2 days after Bolshevik Revolution.
1922	Glavlit, main censorship agency, established
1931 (May)	Ostankino founded, first experimental transmissions.
1939 (10 March)	First regular broadcasting service.
1967 (1 October)	Colour transmissions begin.
1985 (March)	Gorbachev comes to power, policy of *glasnost'* launched.
1990 (1 August)	Media Law abolishes censorship.
1991 (Spring)	Rossiya, Russian Television, second main state channel established.
1991 (18-21 August)	Failed coup by Communist Party hardliners against Gorbachev.
1991 (31 December)	Soviet Union dissolved.
1993 (January)	TV-6, first national private channel, established. Followed later in 1993 by founding of other private channels including NTV.
1993 (October)	Yeltsin's conflict with Russian Parliament ends in assault on latter.
1993 (December)	New Parliament established and new constitution adopted. Federal Television and Radio Broadcasting Agency established.
1994 (November)	Ostankino, Channel 1, main state channel semi-privatised into ORT.
1996 (June)	NTV announces plans for satellite television, Gasprom takes 30% share in company.
1996 (July)	Yeltsin wins Russian Presidential elections.

Abbreviations

GLAVLIT	Main Directorate on Matters of Literature and Publishing. (Chief Censor)
GOSTELERADIO	State Committee of Television and Radio
MNVK	Moscow Independent Broadcasting Corporation (Owner of TV-6)
MTK	Moscow Tele-Radio Broadcasting Company (Owner of Moskva)
NTV	Independent Television
ORT	Russian Public Television Company
RTR	Russian Television and Radio (Owner of Rossiya - see also VGTRK)
TASS	Telegraph Agency of the Soviet Union
VGTRK	All-Russian State Tele-Radio Company (Owner of Rossiya - see also RTR)

Bibliography

Argumenty i fakty, No. 4, January 1993.
Argumenty i fakty, No.12, March 1995.
Benn, David Wedgwood (1992a), '*Glasnost'* and the Media' in S.White, A. Pravda and Z. Gitelman (eds.) *Developments in Soviet and Post-Soviet Politics*, London, Macmillan, pp.174-197.
Benn, David Wedgwood (1992b) *From Glasnost' to Freedom of Speech: Russian Openness and International Relations*, London, Pinter.
Blagovolin, Sergei (1995), 'Vmeshatel'stva konkretnykh lits v veshchanie ORT ne budet', interview with Blagovolin in *Moskovskie novosti*, no.22, 2-9 April, p.21.
Bochkarev, Vidual. (1995), 'Nemnogo obshchikh slov v preddverii zimnego telesezonal' *in Moskovskii komsomolets*, 7 September, p.8.
Bogomolov, Yurii (1995a), 'Muzhskoe i zhenskoe: reiting' in *Moskovskie novosti*, no.13, 19-26 February, p.21.
Bogomolov, Yurii (1995b), 'Peretyagivanie kanalov' in *Moskovskie novosti*, no.17, 12-19 March, p.3.
Fedorovski, V1adimir (1994), 'La corruption reste le problème numéro un de la télévision russe' in *Les ecrits de 1'image*, no.2, Spring 1994, pp.183-187.
McNair, Brian (1991), *Glasnost', Perestroika and the Soviet Media*, London, Routledge.
Mickiewicz, Ellen (1988a), 'Mass Culture, Change and Mobilisation: The Media Revolution' in *Soviet Union*, vol.15, pp.187-200.
Mickiewicz, Ellen (1988b), *Split Signals. Television and Politics in the Soviet Union*, New York, Oxford University Press.
Mickiewicz, Ellen. (1995), 'The Political Economy of Media Democratisation' in D.Lane (ed.) *Russian in Transition: Politics, Privatisation and Inequality*. London, Longman, pp.159-173.
Moscow News, No.23, 13-19 June 1996, p.7.
Moskovskii komsomolets, 20 July 1995, article, 'Regional'noe TV: zakhvat po-amerikanski'.
Moskovskie novosti, no.7, 17-24 Feburary 1991.
Moskovskie novosti, no.31, 30 April-7 May, 1995, p.31, article - 'Reklamnyi rynok: krizis - ne krakh'.
Moskovskie novosti, no.34, 14-21 May, 1995, p.20.
Moskovskie novosti, no.10, 10-17 March, 1996, p.38.
Moskovskie novosti, no.12, 24-31 March, 1996, p.39.
Nezavisimaya gazeta, 10 October 1992, p.5
Nivat, Anne (1996a), 'Russian Politicians Have Mixed Success with Television Campaigning' in *Transition*, 19 April , pp.34-37.
Nivat, Anne (1996b), 'ORT is Governed by No Conception, No Policy' in *Transition*, 19 April, pp.41 and 64.
Rasumovskii, Feliks (1995), 'Esli by zhiv byl apostol Pavel', Interview with Rasumovskii in *Moskovskie novosti*, no.9, 5-12 February, p.22.
Time, 4 March, 1995.
Tolz, Vera (1993), 'The New Role of the Media and Public Opinion under Mikhail Gorbachev' in *The Journal of Communist Studies*, vol.9, no.l, pp.192-209.
Tolz Vera and Wishnevsky Julia (1993) 'The Russian Media and the Political Crisis in Moscow' in *Radio Free Europe/Radio Liberty Research Report*, vol.2, no.40, 8 October, pp.12-15.
Vul'chek, Vsevolod and Muratov, Sergei (1995), 'V telsetyakh TsRU' in *Moskovskie novosti*, no.50, 23-30 July, p.22.
Wishnevsky. Julia (1994) 'The Russian Media after the State of Emergency' in *Radio Free Europe/ Radio Liberty Research Report*, vol.3, no.6, 11 February, pp.1-6.
Yasman, Victor (1988), '*Glasnost'* and the Soviet T.V. - 1987' in *Radio Free Europe/ Radio Liberty Research Report*, 28 January.

2016 x1

2008

14